深度学习系列

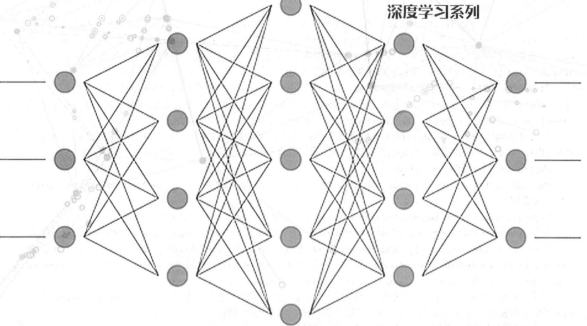

Mastering TensorFlow 1.x

精通 TensorFlow

[美] 阿曼多·凡丹戈（Armando Fandango） 著

刘 波 何希平 译

机械工业出版社

TensorFlow是目前最流行的数值计算库，专用于构建分布式、云计算和移动环境。TensorFlow将数据表示为张量，将计算表示为计算图。

本书是一本综合指南，可让您理解TensorFlow 1.x的高级功能，深入了解TensorFlow 内核、Keras、TF Estimator、TFLearn、TF Slim、PrettyTensor和Sonnet。利用TensorFlow和Keras提供的功能，使用迁移学习、生成对抗网络和深度强化学习等概念来构建深度学习模型。通过本书，您将获得在各种数据集（例如MNIST、CIFAR-10、PTB、text8和COCO图像）上的实践经验。

您还能够学习TensorFlow1.x的高级功能，例如分布式TensorFlow，使用TensorFlow服务部署生产模型，以及在Android和iOS平台上为移动和嵌入式设备构建和部署TensorFlow模型。您将看到如何在R统计软件中调用TensorFlow和Keras API，还能了解在TensorFlow的代码无法按预期工作时所需的调试技术。

本书可帮助您深入了解TensorFlow，使您成为解决人工智能问题的专家。总之，在学习本书之后，可掌握TensorFlow和Keras的产品，并获得构建更智能、更快速、更高效的机器学习和深度学习系统所需的技能。

Copyright © 2018 Packt Publishing

First published in the English language under the title "Mastering TensorFlow 1.x" / by Armando Fandango/ ISBN: 978-1-78829-206-1

Copyright in the Chinese language (simplified characters) © 2018 China Machine Press

This translation of Mastering TensorFlow 1.x first published in 2019 is published by arrangement with Packt Publishing Ltd.

This title is published in China by China Machine Press with license from Packt Publishing Ltd. This edition is authorized for sale in China only, excluding Hong Kong SAR, Macao SAR and Taiwan. Unauthorized export of this edition is a violation of the Copyright Act. Violation of this Law is subject to Civil and Criminal Penalties.

本书由Packt Publishing Ltd授权机械工业出版社在中华人民共和国境内（不包括香港、澳门特别行政区及台湾地区）出版与发行。未经许可的出口，视为违反著作权法，将受法律制裁。

北京市版权局著作权合同登记　图字：01-2018-1732号。

图书在版编目（CIP）数据

精通TensorFlow /（美）阿曼多·凡丹戈（Armando Fandango）著；刘波，何希平译 .—北京：机械工业出版社，2019.1

（深度学习系列）

书名原文：Mastering TensorFlow 1.x

ISBN 978-7-111-61436-4

Ⅰ. ①精… Ⅱ. ①阿… ②刘… ③何… Ⅲ. ①人工智能 – 算法 Ⅳ. ① TP18

中国版本图书馆 CIP 数据核字（2018）第 267302 号

机械工业出版社（北京市百万庄大街 22 号　邮政编码 100037）
策划编辑：刘星宁　　　责任编辑：刘星宁
责任校对：肖　琳　　　责任印制：孙　炜
天津嘉恒印务有限公司印刷
2019 年 1 月第 1 版第 1 次印刷
184mm×240mm · 20.75 印张 · 462 千字
0 001—4 000 册
标准书号：ISBN 978-7-111-61436-4
定价：89.00 元

凡购本书，如有缺页、倒页、脱页，由本社发行部调换
电话服务　　　　　　　　　网络服务
服务咨询热线：010-88361066　机工官网：www.cmpbook.com
读者购书热线：010-68326294　机工官博：weibo.com/cmp1952
　　　　　　　010-88379203　金 书 网：www.golden-book.com
封面无防伪标均为盗版　　　　教育服务网：www.cmpedu.com

译者序

随着人工智能的兴起，深度学习成为人们关注的焦点，也成为机器学习的重要分支。通常所说的深度学习模型是指多层神经网络，它的特点是结构复杂，构建比较困难，训练效率低。

TensorFlow 是 Google 公司于 2015 年 11 月发布的一款深度学习开源框架。它是目前最流行的深度学习框架和机器学习框架。TensorFlow 支持主流的桌面操作系统平台（比如 Linux、Windows 等），也支持主流的嵌入式操作系统（iOS 和 Android），它能有效利用嵌入式设备、台式机、服务器、集群中的 CPU 和 GPU 资源来高效地计算。人们可用 Python、C++ 等多种语言对其进行开发。TensorFlow 不仅提供了非常丰富的深度学习 API，也提供各个矩阵计算方法、最优化算法、可视化工具和调试环境。TensorFlow 还有非常完善的帮助文档。目前在计算机视觉、自然语言处理、推荐系统等领域广泛使用 TensorFlow 作为基础开发平台。

本书用 18 章来介绍怎样使用 TensorFlow 框架。从 TensorFlow 的基本编程入手，介绍了与之相关的核心组件、高级库、计算图和用于查看程序执行过程的 TensorBoard。本书也对如何使用 Keras 创建和训练神经网络模型进行了详细介绍。TensorFlow 可用来实现经典的机器学习模型，因此本书用 1 章来介绍如何用 TensorFlow 实现线性回归、分类等经典机器学习模型。用 TensorFlow 实现深度学习模型会非常方便，这些内容是本书的重点，本书会重点介绍多层感知机（MLP）、循环神经网络（RNN）、卷积神经网络（CNN）、自编码器、深度强化学习、生成对抗网络（GAN）的实现；会通过具体的应用来介绍这些深度学习模型的原理。除了介绍深度学习模型以外，本书还会介绍与训练模型相关的技术，比如迁移学习、分布式训练方法、模型的调试和在生产环境中部署深度学习模型的方法。总之，本书全面涵盖了 TensorFlow 的重要内容，通过对本书的学习，可以掌握 TensorFlow 的编程技巧，可提高工程实战能力，能深入理解各种深度学习模型的基本原理。

本书主要由重庆工商大学人工智能学院刘波博士翻译；重庆工商大学人工智能学院何希平教授对本书的翻译工作进行了技术指导，并负责全书的技术审稿和部分内容的翻译。

翻译本书的过程也是译者不断学习的过程。为了保证专业词汇翻译的准确性，我们在翻译过程中查阅了大量相关资料。但由于时间和能力有限，书中内容难免出现差错。若有问题，读者可通过电子邮件 liubo7971@163.com 与我们联系，欢迎一起探讨，共同进步。

译　者

原书序

我在牛津大学讲授"物联网数据科学"课程时，将 TensorFlow 和 Keras 作为该课程的重要组成部分。我从 Keras 开始接触 TensorFlow，但在我讲授课程时，我越来越倾向于使用 TensorFlow。我相信很多人都会有这样的感受。

本书为学习 TensorFlow 提供了路线图，也会交替介绍 Keras 和 TensorFlow 核心库。本书会深入介绍复杂的主题和相关的库，如 Sonnet、分布式 TensorFlow、使用 TensorFlow 服务部署生产模型、TensorFlow 在移动设备和嵌入式设备上的开发。

本书涉及很多先进的技术，比如深度学习模型中的 RNN、CNN、自编码器、生成对抗模型和深度强化学习。Armando Fandango 利用他的经验清楚地讲解了这些内容，有助于读者理解这些复杂的内容。

我期待更多读者能够阅读本书并从中获益。

Ajit Jaokar
牛津大学物联网课程创建者和首席数据科学家

原书前言

Google公司的TensorFlow已经成为开发人员在应用程序中引入智能处理技术的主要开发工具。与此同时，TensorFlow已经成为每个组织的主要研究工具和工程工具。因此，需要进一步学习TensorFlow的高级用例，这些用例可以在各种软件和设备上构建智能系统。TensorFlow凭借迅速的迭代更新以及代码调试，可将智能理念带入到项目中。因此，精通TensorFlow是创造先进的机器学习、深度学习应用的必由之路。本书将帮您学习TensorFlow的所有高级功能。为了使读者学习进入人工智能世界所需的专业知识，本书汇集了关键信息，因此本书可拓展中级TensorFlow用户的知识面，使其达到更高水平。本书涵盖了从实现先进计算到接近现实世界的研究领域。有了这个非常全面的指南，可以让读者在开发者社区中充分施展才能，也为读者提供了一个为研究工作或项目做出贡献的平台。

本书读者对象

本书适用于任何想要用TensorFlow解决深度学习问题的人员，也适用于那些正在寻找易于理解的技术指南，想深入学习复杂使用案例的技术人员。要想充分利用本书，需要对TensorFlow和Python有基本的了解。

本书涵盖的内容

第1章，TensorFlow 101。本章介绍了TensorFlow的基础知识，例如如何创建张量、常量、变量、占位符和相关操作；了解计算图以及如何将计算图节点放置在各种设备（如GPU）上。通过本章，还可以学习如何使用TensorBoard来可视化各种中间结果和最终的输出值。

第2章，TensorFlow的高级库。本章介绍了几个高级库，包括TF Contrib Learn、TF Slim、TFLearn、Sonnet和Pretty Tensor等。

第3章，Keras 101。本章详细介绍了高级库Keras，这是TensorFlow核心之一。

第4章，基于TensorFlow的经典机器学习算法。本章介绍如何使用TensorFlow来实现经典的机器学习算法，如线性回归和逻辑回归分类。

第5章，基于TensorFlow和Keras的神经网络和多层感知机。本章介绍了神经网络的概念，并介绍如何构建简单的神经网络模型，还介绍了如何构建多层感知机（MLP）。

第6章，基于TensorFlow和Keras的RNN。本章介绍如何用TensorFlow和Keras构建RNN。这里将介绍RNN、长短期记忆（LSTM）网络和门控循环单元（GRU）网络的内部结构。本章还会简要介绍在TensorFlow和Keras中用于实现RNN模型的API函数和类。

第7章，基于TensorFlow和Keras的RNN在时间序列数据中的应用。本章介绍了如何构建和训练基于时间序列数据的RNN模型，并提供相应的示例。

第8章，基于TensorFlow和Keras的RNN在文本数据中的应用。本章介绍如何构建和训练基于文本数据的 RNN 模型，并给出相应的示例。这一章会学习使用 TensorFlow 和 Keras 来构建词向量，然后通过基于词向量嵌入的 LSTM 模型从示例文本数据中生成文本。

第 9 章，基于 TensorFlow 和 Keras 的 CNN。本章介绍用于处理图像数据的 CNN 模型，并给出基于 TensorFlow 和 Keras 库的示例，该示例实现了 LeNet 架构模式。

第 10 章，基于 TensorFlow 和 Keras 的自编码器。本章介绍了用于处理图像数据的自编码模型，并给出基于 TensorFlow 和 Keras 库的示例，同时也给出了简单自编码器、去噪自编码器和变分自编码器的实现。

第 11 章，使用 TF 服务提供生成环境下的 TensorFlow 模型。本章会介绍如何用 TensorFlow 服务来部署模型，学习如何在 Docker 容器和 Kubernetes 集群中使用 TF 服务进行部署。

第 12 章，迁移学习模型和预训练模型。本章会介绍如何用预训练模型来进行预测，学习如何在不同的数据集上重新训练模型。本章提供基于 VGG16 和 Inception V3 模型的示例，这些模型预先在 ImageNet 数据集上训练好，并用来预测 COCO 数据集中的图像；也会介绍一个示例，它通过使用 COCO 数据集重新训练模型的最后一层来改进预测效果。

第 13 章，深度强化学习。本章会介绍强化学习和 OpenAIgym 框架。也会介绍如何使用各种强化学习策略（包括深度 Q 网络）来构建和训练多个模型。

第 14 章，生成对抗网络（GAN）。本章介绍如何通过 TensorFlow 和 Keras 来构建和训练生成对抗模型。

第 15 章，基于 TensorFlow 集群的分布式模型。本章将介绍如何基于 TensorFlow 集群来训练 TensorFlow 模型，这是一种分布式训练方法。本章还会通过示例介绍以数据并行方式训练模型时的异步和同步更新方法。

第 16 章，移动和嵌入式平台上的 TensorFlow 模型。本章会介绍如何在基于 iOS 和 Android 平台的移动设备上部署 TensorFlow 模型。本章涵盖了基于 TensorFlow 库的 TF Mobile 和 TF Lite API。

第 17 章，R 中的 TensorFlow 和 Keras。本章介绍了如何在 R 统计软件中构建和训练 TensorFlow 模型，还介绍 R Studio 用于实现 TF Core，TF Estimators 和 Keras API 的三个软件包。

第 18 章，调试 TensorFlow 模型。本章会介绍当模型无法按预期工作时，如何发现热点问题的策略和技巧。本章涵盖了 TensorFlow 调试器和其他调试方法。

附录，张量处理单元。本章简要介绍了张量处理单元（TPU）。TPU 是用来训练和运行 TensorFlow 模型的重要平台，它能优化整个处理过程。虽然 TensorFlow 尚未广泛使用，但已经被用于 Google 云平台（Google Cloud Platform，GCP），并且很快将在 GCP 之外推广。

充分利用本书

1）我们假设您熟悉 Python 编程并具有 TensorFlow 和 Keras 的基础知识。
2）如果您具有这方面的知识，请安装 Jupyter 笔记本、TensorFlow 和 Keras。
3）下载本书包含的 Python、R 和 notebook 代码文件。
4）在阅读本书时需要运行这些代码，并通过修改提供的示例代码来进行深度学习。

5）若要运行与 Android 平台相关的代码，需要有 Android Studio 框架和 Android 设备。

6）若要运行与 iOS 平台相关的代码，需要有 Xcode 和 Apple 设备。

7）在有些与 TensorFlow 相关的章节，需要安装 Docker 和 Kubernetes。本书提供了在 Ubuntu 上安装 Kubernetes 和 Docker 的说明。

下载示例代码文件

可以登录 www.packtpub.com 来下载本书的示例代码。如果在其他地方购买了本书，也可以访问 www.packtpub.com/support 并注册，我们会直接将文件发给您。

可以按照以下步骤下载代码文件：

1）登录或注册 www.packtpub.com。

2）选择 SUPPORT 选项卡。

3）点击 Code Downloads & Errata。

4）在搜索框中输入图书的名称，然后按照屏幕上的说明进行操作。

下载文件后，请确保使用下面最新版本的解压缩软件来进行解压：

- WinRAR/7-Zip 适用于 Windows。
- Zipeg/iZip/UnRarX 适用于 Mac。
- 7-Zip/PeaZip 适用于 Linux。

本书的代码包也托管在 GitHub 上，具体地址为：https://github.com/PacktPublishing/Mastering-TensorFlow-1x。另外在 https://github.com/PacktPublishing/ 上提供了各类书籍和视频。读者可以试着去看一下！

使用约定

本书使用了许多文本约定。

CodeInText：指文本、数据库表名、文件夹名称、文件名、文件扩展名、路径名、虚拟 URL、用户输入和 Twitter 句柄中的代码字。

例如："with</kbd> block, which will be shown later in this chapter"

代码块会进行如下设置：

```
from datasetslib.ptb import PTBSimple
ptb = PTBSimple()
ptb.load_data()
print('Train :',ptb.part['train'][0:5])
print('Test: ',ptb.part['test'][0:5])
print('Valid: ',ptb.part['valid'][0:5])
print('Vocabulary Length = ',ptb.vocab_len)
```

粗体：表示新术语，重要单词或需要注意的单词。例如，菜单或对话框中的文字就会采用这种格式。例如："从 Administration 面板中选择 System info。"

警告或重要提示在此显示。

提示和技巧在此显示。

联系我们

欢迎读者反馈。

一般反馈：可发送电子邮件到 feedback@packtpub.com，并在邮件主题中写上书名。如果您对本书的任何方面有疑问，请发送电子邮件到 questions@packtpub.com。

勘误表：尽管我们已经尽全力确保内容的准确性，但仍会发生错误。如果您在本书中发现错误，请告诉我们，我们将不胜感激。请访问 www.packtpub.com/submit-errata，选择您的书籍，单击勘误提交表单链接，然后输入详细信息。

盗版：如果读者在互联网上发现任何非法复制我们作品的情况，请将网址或网站名称提供给我们，我们将不胜感激。请通过 copyright@packtpub.com 与我们联系并提供材料链接。

成为作者：如果您有专业的知识，并且对撰写书籍感兴趣，请访问 authors.packtpub.com。

评论

请留下您的评论。一旦您阅读并使用了本书，为什么不在购买的网站上留下评论呢？潜在的读者可以看到并根据您中肯的意见做出购买决定，我们在 Packt 网站上可以了解您对我们产品的看法。同时，我们的作者可以看到您对他们著作的反馈。谢谢您！

有关 Packt 的更多信息，请访问 packtpub.com。

目　录

译者序

原书序

原书前言

第1章　TensorFlow 101 // 1
1.1　什么是 TensorFlow // 1
1.2　TensorFlow 内核 // 2
 1.2.1　简单的示例代码 -Hello TensorFlow // 2
 1.2.2　张量 // 3
 1.2.3　常量 // 4
 1.2.4　操作 // 5
 1.2.5　占位符 // 6
 1.2.6　从 Python 对象创建张量 // 7
 1.2.7　变量 // 9
 1.2.8　由库函数生成的张量 // 10
 1.2.9　通过 tf.get_variable( ) 获取变量 // 13
1.3　数据流图或计算图 // 14
 1.3.1　执行顺序和延迟加载 // 15
 1.3.2　跨计算设备执行计算图 -CPU 和 GPU // 15
 1.3.3　多个计算图 // 18
1.4　TensorBoard // 19
 1.4.1　TensorBoard 最小的例子 // 19
 1.4.2　TensorBoard 的细节 // 21
1.5　总结 // 21

第2章　TensorFlow 的高级库 // 22
2.1　TF Estimator // 22
2.2　TF Slim // 24
2.3　TFLearn // 25
 2.3.1　创建 TFLearn 层 // 26
 2.3.2　创建 TFLearn 模型 // 30
 2.3.3　训练 TFLearn 模型 // 30
 2.3.4　使用 TFLearn 模型 // 30
2.4　PrettyTensor // 31
2.5　Sonnet // 32
2.6　总结 // 34

第3章　Keras101 // 35
3.1　安装 Keras // 35
3.2　Keras 的神经网络模型 // 36
 3.2.1　在 Keras 中创建模型的过程 // 36
3.3　创建 Keras 模型 // 36
 3.3.1　用于创建 Keras 模型的序列化 API // 36
 3.3.2　用于创建 Keras 模型的功能性 API // 37
3.4　Keras 的层 // 37
 3.4.1　Keras 内核层 // 37
 3.4.2　Keras 卷积层 // 38
 3.4.3　Keras 池化层 // 38
 3.4.4　Keras 局连接层 // 39
 3.4.5　Keras 循环层 // 39
 3.4.6　Keras 嵌入层 // 39

3.4.7 Keras 合并层 // 39
3.4.8 Keras 高级激活层 // 40
3.4.9 Keras 归一化层 // 40
3.4.10 Keras 噪声层 // 40
3.5 将网络层添加到 Keras 模型中 // 40
　3.5.1 利用序列化 API 将网络层添加到 Keras 模型中 // 40
　3.5.2 利用功能性 API 将网络层添加到 Keras 模型中 // 41
3.6 编译 Keras 模型 // 41
3.7 训练 Keras 模型 // 42
3.8 使用 Keras 模型进行预测 // 42
3.9 Keras 中的其他模块 // 43
3.10 基于 MNIST 数据集的 Keras 顺序模型示例 // 43
3.11 总结 // 45

第 4 章 基于 TensorFlow 的经典机器学习算法 // 47

4.1 简单的线性回归 // 48
　4.1.1 数据准备 // 49
　4.1.2 建立简单的回归模型 // 50
　4.1.3 使用训练好的模型进行预测 // 55
4.2 多元回归 // 55
4.3 正则化回归 // 58
　4.3.1 Lasso 正则化 // 59
　4.3.2 岭正则化 // 62
　4.3.3 弹性网正则化 // 64
4.4 使用 Logistic 回归进行分类 // 65
　4.4.1 二分类的 Logistic 回归 // 65
　4.4.2 多类分类的 Logistic 回归 // 66
4.5 二分类 // 66
4.6 多分类 // 69

4.7 总结 // 73

第 5 章 基于 TensorFlow 和 Keras 的神经网络和多层感知机 // 74

5.1 感知机 // 74
5.2 多层感知机 // 76
5.3 用于图像分类的多层感知机 // 77
　5.3.1 通过 TensorFlow 构建用于 MNIST 分类的多层感知机 // 77
　5.3.2 通过 Keras 构建用于 MNIST 分类的多层感知机 // 83
　5.3.3 通过 TFLearn 构建用于 MNIST 分类的多层感知机 // 85
　5.3.4 多层感知机与 TensorFlow、Keras 和 TFLearn 的总结 // 86
5.4 用于时间序列回归的多层感知机 // 86
5.5 总结 // 89

第 6 章 基于 TensorFlow 和 Keras 的 RNN // 90

6.1 简单 RNN // 90
6.2 RNN 改进版本 // 92
6.3 LSTM 网络 // 93
6.4 GRU 网络 // 95
6.5 基于 TensorFlow 的 RNN // 96
　6.5.1 TensorFlow 的 RNN 单元类 // 96
　6.5.2 TensorFlow 的 RNN 模型构造类 // 97
　6.5.3 TensorFlow 的 RNN 单元封装类 // 97
6.6 基于 Keras 的 RNN // 98
6.7 RNN 的应用领域 // 98
6.8 将基于 Keras 的 RNN 用于 MNIST 数据 // 99

6.9 总结 // 100

第 7 章 基于 TensorFlow 和 Keras 的 RNN 在时间序列数据中的应用 //101

7.1 航空公司乘客数据集 // 101
    7.1.1 加载 airpass 数据集 // 102
    7.1.2 可视化 airpass 数据集 // 102

7.2 使用 TensorFlow 为 RNN 模型预处理数据集 // 103

7.3 TensorFlow 中的简单 RNN // 104

7.4 TensorFlow 中的 LSTM 网络 // 106

7.5 TensorFlow 中的 GRU 网络 // 107

7.6 使用 Keras 为 RNN 模型预处理数据集 // 108

7.7 基于 Keras 的简单 RNN // 109

7.8 基于 Keras 的 LSTM 网络 // 111

7.9 基于 Keras 的 GRU 网络 // 112

7.10 总结 // 113

第 8 章 基于 TensorFlow 和 Keras 的 RNN 在文本数据中的应用 // 114

8.1 词向量表示 // 114

8.2 为 word2vec 模型准备数据 // 116
    8.2.1 加载和准备 PTB 数据集 // 117
    8.2.2 加载和准备 text8 数据集 // 118
    8.2.3 准备小的验证集 // 119

8.3 使用 TensorFlow 的 skip-gram 模型 // 119

8.4 使用 t-SNE 可视化单词嵌入 // 124

8.5 基于 Keras 的 skip-gram 模型 // 126

8.6 使用 TensorFlow 和 Keras 中的 RNN 模型生成文本 // 130

    8.6.1 使用 TensorFlow 中的 LSTM 模型生成文本 // 131
    8.6.2 使用 Keras 中的 LSTM 模型生成文本 // 134

8.7 总结 // 137

第 9 章 基于 TensorFlow 和 Keras 的 CNN // 138

9.1 理解卷积 // 138

9.2 理解池化 // 141

9.3 CNN 架构模式 - LeNet // 142

9.4 在 MNIST 数据集上构建 LeNet // 143
    9.4.1 使用 TensorFlow 的 LeNet CNN 对 MNIST 数据集进行分类 // 143
    9.4.2 使用 Keras 的 LeNet CNN 对 MNIST 数据集进行分类 // 146

9.5 在 CIFAR10 数据集上构建 LeNet // 148
    9.5.1 使用 TensorFlow 的 CNN 对 CIFAR10 数据集进行分类 // 149
    9.5.2 使用 Keras 的 CNN 对 CIFAR10 数据集进行分类 // 150

9.6 总结 // 151

第 10 章 基于 TensorFlow 和 Keras 的自编码器 // 152

10.1 自编码器类型 // 152

10.2 基于 TensorFlow 的堆叠自编码器 // 154

10.3 基于 Keras 的堆叠自编码器 // 157

10.4 基于 TensorFlow 的去噪自编码器 // 159

10.5 基于 Keras 的去噪自编码器 // 161

10.6 基于 TensorFlow 的变分自编码器 // 162

10.7 基于 Keras 的变分自编码器 // 167

10.8 总结 // 170

第 11 章 使用 TF 服务提供生成环境下的 TensorFlow 模型 // 171

11.1 在 TensorFlow 中保存和恢复模型 // 171

    11.1.1 使用 saver 类保存和恢复所有网络计算图变量 // 172

    11.1.2 使用 saver 类保存和恢复所选变量 // 173

11.2 保存和恢复 Keras 模型 // 175

11.3 TensorFlow 服务 // 175

    11.3.1 安装 TF 服务 // 175

    11.3.2 保存 TF 服务的模型 // 176

    11.3.3 使用 TF 服务提供服务模型 // 180

11.4 在 Docker 容器中提供 TF 服务 // 181

    11.4.1 安装 Docker // 182

    11.4.2 为 TF 服务构建 Docker 镜像 // 183

    11.4.3 在 Docker 容器中提供模型 // 185

11.5 基于 Kubernetes 的 TF 服务 // 186

    11.5.1 安装 Kubernetes // 186

    11.5.2 将 Docker 镜像上传到 dockerhub // 187

    11.5.3 在 Kubernetes 中部署 // 188

11.6 总结 // 192

第 12 章 迁移学习模型和预训练模型 // 193

12.1 ImageNet 数据集 // 193

12.2 重新训练或微调模型 // 196

12.3 COCO 动物数据集和预处理图像 // 197

12.4 TensorFlow 中的 VGG16 // 203

    12.4.1 使用 TensorFlow 中预先训练的 VGG16 进行图像分类 // 204

12.5 将 TensorFlow 中的图像预处理用于预先训练的 VGG16 // 208

    12.5.1 使用 TensorFlow 中重新训练的 VGG16 进行图像分类 // 209

12.6 Keras 中的 VGG16 // 215

    12.6.1 使用 Keras 中预先训练的 VGG16 进行图像分类 // 215

    12.6.2 使用 Keras 中重新训练的 VGG16 进行图像分类 // 220

12.7 TensorFlow 中的 Inception v3 // 226

    12.7.1 使用 TensorFlow 中 Inception v3 进行图像分类 // 226

    12.7.2 使用 TensorFlow 中重新训练的 Inception v3 进行图像分类 // 231

12.8 总结 // 237

第 13 章 深度强化学习 // 238

13.1 OpenAI Gym 101 // 239

13.2 将简单的策略应用于 cartpole 游戏 // 242

13.3 强化学习 101 // 246

    13.3.1 Q 函数（在模型无效时学习优化）// 246

    13.3.2 强化学习算法的探索与开发 // 246

    13.3.3 V 函数（在模型可用时学习优化）// 247

    13.3.4 强化学习技巧 // 247

13.4 强化学习的朴素神经网络策略 // 248

13.5 实施 Q-Learning // 250

    13.5.1 Q-Learning 的初始化和离散化 // 251

13.5.2 基于 Q 表的 Q-Learning // 252

13.5.3 使用 Q 网络或深度 Q 网络（DQN）进行 Q-Learning // 253

13.6 总结 // 254

第 14 章 生成对抗网络（GAN）// 256

14.1 GAN 101 // 256

14.2 建立和训练 GAN 的最佳实践 // 258

14.3 基于 TensorFlow 的简单 GAN // 258

14.4 基于 Keras 的简单 GAN // 263

14.5 基于 TensorFlow 和 Keras 的深度卷积 GAN // 268

14.6 总结 // 270

第 15 章 基于 TensorFlow 集群的分布式模型 // 271

15.1 分布式执行策略 // 271

15.2 TensorFlow 集群 // 272

15.2.1 定义集群规范 // 274

15.2.2 创建服务器实例 // 274

15.2.3 定义服务器和设备之间的参数和操作 // 276

15.2.4 定义并训练计算图以进行异步更新 // 276

15.2.5 定义并训练计算图以进行同步更新 // 281

15.3 总结 // 282

第 16 章 移动和嵌入式平台上的 TensorFlow 模型 // 283

16.1 移动平台上的 TensorFlow // 283

16.2 Android 应用程序中的 TF Mobile // 284

16.3 演示 Android 上的 TF Mobile // 285

16.4 iOS 应用程序中的 TF Mobile // 287

16.5 演示 iOS 上的 TF Mobile // 288

16.6 TensorFlow Lite // 289

16.7 演示 Android 上的 TF Lite 应用程序 // 290

16.8 演示 iOS 上的 TF Lite 应用程序 // 291

16.9 总结 // 291

第 17 章 R 中的 TensorFlow 和 Keras // 292

17.1 在 R 中安装 TensorFlow 和 Keras 软件包 // 292

17.2 R 中的 TF 核心 API // 294

17.3 R 中的 TF Estimator API // 295

17.4 R 中的 Keras API // 297

17.5 R 中的 TensorBoard // 300

17.6 R 中的 tfruns 包 // 302

17.7 总结 // 304

第 18 章 调试 TensorFlow 模型 // 305

18.1 使用 tf.Session.run( ) 获取张量值 // 305

18.2 使用 tf.Print( ) 输出张量值 // 306

18.3 使用 tf.Assert( ) 断言条件 // 306

18.4 使用 TensorFlow 调试器（tfdbg）进行调试 // 308

18.5 总结 // 310

附录 张量处理单元 // 311

第 1 章
TensorFlow 101

TensorFlow 是一种用来解决机器学习和深度学习问题的流行库，由 Google 开发，在内部使用之后，被作为开源项目公开，并继续进行开发和使用。本章将介绍 TensorFlow 的三种模型：数据模型、编程模型和执行模型。

TensorFlow 的数据模型由张量组成；编程模型由数据流图或计算图组成；执行模型由系列基于依赖条件的触发（firing）节点组成，从初始节点（它与输入数据有关）开始执行。

本章将介绍构成这三种模型的 TensorFlow 的要素，也称为 TensorFlow 的内核。

本章将介绍以下主题：

- TensorFlow 内核：
  - 张量；
  - 常量；
  - 操作；
  - 占位符；
  - 从 Python 对象创建张量；
  - 变量；
  - 由库函数生成张量。
- 数据流图或计算图：
  - 执行顺序和延迟加载；
  - 跨计算设备执行计算图 -CPU 和 GPU；
  - 多个计算图。
- TensorBoard。

> 本书注重编程实践，因此可从 GitHub 复制或从 Packt Publishing 网站下载本书的程序代码。可以在代码包 Jupyter 笔记本 ch-01_TensorFlow_101 中查看本章的示例代码。

1.1 什么是 TensorFlow

TensorFlow 官网（www.tensorflow.org）是这样解释的：

> TensorFlow 是一个使用数据流图进行数值计算的开源库。

TensorFlow 最初由 Google 为其内部使用而开发，于 2015 年 11 月 9 日以开源形式发布。从那时起，TensorFlow 已广泛应用于开发各个领域的机器学习和深度神经网络模型，并继续在 Google 内部用于研究和产品开发。TensorFlow 1.0 于 2017 年 2 月 15 日发布，这让人联想到 TensorFlow 是 Google 特意送给机器学习工程师的情人节礼物！

TensorFlow 包括数据模型、编程模型和执行模型：

- **数据模型**由张量组成，这些张量是 TensorFlow 程序进行创建、处理和保存的基本数据单元。
- **编程模型**包含数据流图或计算图，在 TensorFlow 中创建程序意味着构建一个或多个 TensorFlow 计算图。
- **执行模型**由一系列相互依赖的计算图的节点组成，从直接连接输入的节点开始执行，并仅依赖于当前的输入。

要在您的项目中使用 TensorFlow，需要学习如何使用 TensorFlow API 进行编程。TensorFlow 有多个能与库进行交互的 API。TF 的 API 或库分为两个级别：

- **底层库**：底层库也称为 TensorFlow 内核，它提供了非常全面（fine-grained）的底层功能，从而可以完全掌握如何在模型中使用库。本章将介绍 TensorFlow 内核。
- **高级库**：高级库提供了高级功能，能够相对容易地实现模型。这些库包括 TF Estimators、TFLearn、TFSlim、Sonnet 和 Keras。下一章将会介绍其中的一些库。

1.2 TensorFlow 内核

TensorFlow 内核是构建更高级别 TensorFlow 模块的低级库，在深入学习高级 TensorFlow 之前，了解底层库的概念非常重要。本节将介绍所有底层库的核心概念。

1.2.1 简单的示例代码 -Hello TensorFlow

学习任何新的编程语言、库或平台时，往往应先学习编写简单的 Hello TensorFlow 代码。

假设已经安装了 TensorFlow，如果还没有，请参阅 TensorFlow 的安装指南：https://www.tensorflow.org/install/。这里有关于安装 TensorFlow 的详细说明。

在 Jupyter 笔记本中打开文件 ch-01_TensorFlow_101.ipynb，通过运行下面的代码来学习。

1）通过以下代码导入 TensorFlow 库：

```
import tensorflow as tf
```

2）创建 TensorFlow 会话。TensorFlow 提供两种会话：Session( ) 和 InteractiveSession( )。使用以下代码创建交互式会话：

```
tfs = tf.InteractiveSession()
```

> Session() 和 InteractiveSession() 之间的唯一区别是用 InteractiveSession() 创建的会话为默认会话，因此，不需要指定会话上下文来确定稍后要执行的命令（这些命令与会话相关）。例如，有一个会话对象 tfs 和一个常量对象 hello。如果 tfs 是 InteractiveSession() 对象，那么可以使用代码 hello.eval() 来执行 hello。如果 tfs 是一个 Session() 对象，那么必须通过 tfs.hello.eval()（或 with 块）来执行。在实际应用中常使用 with 块（本章稍后将进行介绍）。

3）定义一个 TensorFlow 常量 hello：

```
hello = tf.constant("Hello TensorFlow !!")
```

4）在 TensorFlow 会话中执行常量并打印输出：

```
print(tfs.run(hello))
```

5）将得到如下结果：

```
'Hello TensorFlow !!'
```

现在已经使用 TensorFlow 编写并执行了前两行代码，下面来看看 TensorFlow 的基本组成部分。

1.2.2 张量

张量是 TensorFlow 进行计算的基本元素和基本数据结构。可能是学习 TensorFlow 需要了解的唯一数据结构。张量是一个 n 维数据集合，由秩（rank）、形状和类型来标识。

秩是张量的维数，**形状**表示每个维度大小的列表。张量可以有任意数量的维度。大家所熟悉的张量形状有：零维集合（标量）、一维集合（矢量）、二维集合（矩阵）和多维集合。

标量是秩为 0 的张量，因此具有 [1] 的形状。矢量或一维数组是秩为 1 的张量，并具有 [列] 或 [行] 的形状。矩阵或二维数组是秩为 2 的张量，并具有 [行，列] 的形状。三维矩阵是秩为 3 的张量，类似地，n 维矩阵是秩为 n 的张量。

请参阅以下资源以了解有关张量及其数学基础的更多信息：

- 维基百科上的张量页面：https://en.wikipedia.org/wiki/Tensor。
- 美国国家航空航天局介绍张量的用户指南：https://www.grc.nasa.gov/www/k-12/Numbers/Math/documents/Tensors_TM2002211716.pdf。

张量在所有维度上的数据类型要一样，其元素的数据类型被称为张量的数据类型。

> 也可以通过 https://www.tensorflow.org/api_docs/python/tf/DType 来检查最新版本的 TensorFlow 库中所定义的数据类型。

在编写本书时，TensorFlow 定义的数据类型见表 1-1。

表 1-1

TensorFlow Python API 数据类型	描述
tf.float16	16 位半精度浮点
tf.float32	32 位单精度浮点
tf.float64	64 位双精度浮点
tf.bfloat16	16 位截断浮点
tf.complex64	64 位单精度复合体
tf.complex128	128 位双精度复合体
tf.int8	8 位有符号整数
tf.uint8	8 位无符号整数
tf.uint16	16 位无符号整数
tf.int16	16 位有符号整数
tf.int32	32 位有符号整数
tf.int64	64 位有符号整数
tf.bool	布尔类型
tf.string	字符串
tf.qint8	量化的 8 位有符号整数
tf.quint8	量化的 8 位无符号整数
tf.qint16	量化的 16 位有符号整数
tf.quint16	量化的 16 位无符号整数
tf.qint32	量化的 32 位有符号整数
tf.resource	处理一个可变资源

建议避免使用 Python 的本地（native）数据类型，而是尽量使用 TensorFlow 数据类型来定义张量。

可以通过以下方式创建张量：
- 通过定义常量、操作和变量，并将值传递给构造函数。
- 通过定义占位符（placeholder）并将值传递给 session.run( )。
- 通过使用 tf.convert_to_tensor( ) 函数来将标量值、列表和 NumPy 数组等 Python 对象转换成张量。

下面来看看创建张量的不同方法。

1.2.3 常量

常量型张量可按下面所定义的 tf.constant( ) 函数来创建：

```
tf.constant(
  value,
  dtype=None,
  shape=None,
  name='Const',
  verify_shape=False
)
```

来看一下 Jupyter 笔记本提供的示例代码：

```
c1=tf.constant(5,name='x')
c2=tf.constant(6.0,name='y')
c3=tf.constant(7.0,tf.float32,name='z')
```

仔细查看代码：

- 第一行定义了一个常量张量 c1，给它赋值 5，并将其命名为 x。
- 第二行定义了一个常量张量 c2，存储值为 6.0，并将其命名为 y。
- 当打印这些张量时，可以看到 c1 和 c2 的数据类型，这些类型会由 TensorFlow 自动推导出来。
- 对于自定义数据类型，可以使用 dtype 参数或将数据类型作为第二个参数。在前面的代码示例中，c3 的数据类型为 tf.float32。

打印常量 c1、c2 和 c3：

```
print('c1 (x): ',c1)
print('c2 (y): ',c2)
print('c3 (z): ',c3)
```

当打印这些常量时，得到以下输出：

```
c1 (x):  Tensor("x:0", shape=(), dtype=int32)
c2 (y):  Tensor("y:0", shape=(), dtype=float32)
c3 (z):  Tensor("z:0", shape=(), dtype=float32)
```

为了打印这些常量的值，必须在 TensorFlow 会话中执行 tfs.run( ) 命令：

```
print('run([c1,c2,c3]) : ',tfs.run([c1,c2,c3]))
```

可得到以下结果：

```
run([c1,c2,c3]) :  [5, 6.0, 7.0]
```

1.2.4 操作

TensorFlow 提供了许多可应用于张量的操作，可通过传递值⊖并将输出分配给另一个张量来定义操作。例如，在本章的 Jupyter 笔记本文件中，定义了 op1 和 op2 操作：

```
op1 = tf.add(c2,c3)
op2 = tf.multiply(c2,c3)
```

当输出 op1 和 op2 时，发现 op1 和 op2 被定义为张量：

```
print('op1 : ', op1)
print('op2 : ', op2)
```

输出结果如下：

```
op1 :  Tensor("Add:0", shape=(), dtype=float32)
op2 :  Tensor("Mul:0", shape=(), dtype=float32)
```

要输出这些操作的值，必须在 TensorFlow 的会话窗口中运行这些操作：

⊖ 传递给函数的参数值（或张量）。

```
print('run(op1) : ', tfs.run(op1))
print('run(op2) : ', tfs.run(op2))
```

输出结果如下：

```
run(op1) :  13.0
run(op2) :  42.0
```

表 1-2 列出了一些内置操作。

表 1-2

操作类型	操 作
算术运算	tf.add, tf.subtract, tf.multiply, tf.scalar_mul, tf.div, tf.divide, tf.truediv, tf.floordiv, tf.realdiv, tf.truncatediv, tf.floor_div, tf.truncatemod, tf.floormod, tf.mod, tf.cross
基本的数学操作	tf.add_n, tf.abs, tf.negative, tf.sign, tf.reciprocal, tf.square, tf.round, tf.sqrt, tf.rsqrt, tf.pow, tf.exp, tf.expm1, tf.log, tf.log1p, tf.ceil, tf.floor, tf.maximum, tf.minimum, tf.cos, tf.sin, tf.lbeta, tf.tan, tf.acos, tf.asin, tf.atan, tf.lgamma, tf.digamma, tf.erf, tf.erfc, tf.igamma, tf.squared_difference, tf.igammac, tf.zeta, tf.polygamma, tf.betainc, tf.rint
矩阵运算	tf.diag, tf.diag_part, tf.trace, tf.transpose, tf.eye, tf.matrix_diag, tf.matrix_diag_part, tf.matrix_band_part, tf.matrix_set_diag, tf.matrix_transpose, tf.matmul, tf.norm, tf.matrix_determinant, tf.matrix_inverse, tf.cholesky, tf.cholesky_solve, tf.matrix_solve, tf.matrix_triangular_solve, tf.matrix_solve_ls, tf.qr, tf.self_adjoint_eig, tf.self_adjoint_eigvals, tf.svd
张量运算	tf.tensordot
复杂的数值操作	tf.complex, tf.conj, tf.imag, tf.real
字符串操作	tf.string_to_hash_bucket_fast, tf.string_to_hash_bucket_strong, tf.as_string, tf.encode_base64, tf.decode_base64, tf.reduce_join, tf.string_join, tf.string_split, tf.substr, tf.string_to_hash_bucket

1.2.5 占位符

尽管在定义常量张量时可提供一个值，但占位符允许创建可在运行时提供值的张量。TensorFlow 的 tf.placeholder( ) 函数可用来创建占位符张量，该函数的具体定义如下：

```
tf.placeholder(
    dtype,
    shape=None,
    name=None
    )
```

下面这个例子创建了两个占位符张量并输出它们的信息：

```
p1 = tf.placeholder(tf.float32)
p2 = tf.placeholder(tf.float32)
print('p1 : ', p1)
print('p2 : ', p2)
```

可以看到以下输出：

```
p1:Tensor("Placeholder:0", dtype=float32)
p2:Tensor("Placeholder_1:0", dtype=float32)
```

现在来定义一个使用这些占位符的操作：

```
op4 = p1 * p2
```

TensorFlow 允许为各种操作使用简写符号，在前面的例子中，p1 * p2 是 tf.multiply(p1, p2) 的简写形式：

```
print('run(op4,{p1:2.0, p2:3.0}) : ',tfs.run(op4,{p1:2.0, p2:3.0}))
```

上述命令在 TensorFlow 会话中运行 op4，将 p1 和 p2 的值提供给 Python 字典 [run( ) 操作的第二个参数]。

输出结果如下：

```
run(op4,{p1:2.0, p2:3.0}) :  6.0
```

还可以使用 run( ) 操作中的 feed_dict 参数指定字典：

```
print('run(op4,feed_dict = {p1:3.0, p2:4.0}) : ',
      tfs.run(op4, feed_dict={p1: 3.0, p2: 4.0}))
```

输出结果如下：

```
run(op4,feed_dict = {p1:3.0, p2:4.0}) :  12.0
```

来看看最后一个例子，将向量传递给 p1 和 p2：

```
print('run(op4,feed_dict = {p1:[2.0,3.0,4.0], p2:[3.0,4.0,5.0]}) : ',
      tfs.run(op4,feed_dict = {p1:[2.0,3.0,4.0], p2:[3.0,4.0,5.0]}))
```

输出结果如下：

```
run(op4,feed_dict={p1:[2.0,3.0,4.0],p2:[3.0,4.0,5.0]}):[  6.  12.  20.]
```

两个输入向量按逐元素方式相乘。

1.2.6 从 Python 对象创建张量

可以使用 tf.convert_to_tensor( ) 函数从 Python 对象（如列表和 NumPy 数组）创建张量，该函数的定义如下：

```
tf.convert_to_tensor(
  value,
  dtype=None,
  name=None,
  preferred_dtype=None
)
```

下面的代码会创建一些张量并输出相应的结果：

1）创建并输出零维张量：

```
tf_t=tf.convert_to_tensor(5.0,dtype=tf.float64)

print('tf_t : ',tf_t)
print('run(tf_t) : ',tfs.run(tf_t))
```

输出结果如下：

```
tf_t : Tensor("Const_1:0", shape=(), dtype=float64)
run(tf_t) : 5.0
```

2）创建并输出一维张量：

```
a1dim = np.array([1,2,3,4,5.99])
print("a1dim Shape : ",a1dim.shape)

tf_t=tf.convert_to_tensor(a1dim,dtype=tf.float64)

print('tf_t : ',tf_t)
print('tf_t[0] : ',tf_t[0])
print('tf_t[0] : ',tf_t[2])
print('run(tf_t) : \n',tfs.run(tf_t))
```

输出结果如下：

```
a1dim Shape :  (5,)
tf_t : Tensor("Const_2:0", shape=(5,), dtype=float64)
tf_t[0] : Tensor("strided_slice:0", shape=(), dtype=float64)
tf_t[0] : Tensor("strided_slice_1:0", shape=(), dtype=float64)
run(tf_t) :
 [ 1.   2.   3.   4.   5.99]
```

3）创建并输出二维张量：

```
a2dim = np.array([(1,2,3,4,5.99),
                  (2,3,4,5,6.99),
                  (3,4,5,6,7.99)
                  ])
print("a2dim Shape : ",a2dim.shape)

tf_t=tf.convert_to_tensor(a2dim,dtype=tf.float64)

print('tf_t : ',tf_t)
print('tf_t[0][0] : ',tf_t[0][0])
print('tf_t[1][2] : ',tf_t[1][2])
print('run(tf_t) : \n',tfs.run(tf_t))
```

输出结果如下：

```
a2dim Shape :  (3, 5)
tf_t : Tensor("Const_3:0", shape=(3, 5), dtype=float64)
tf_t[0][0] : Tensor("strided_slice_3:0", shape=(), dtype=float64)
tf_t[1][2] : Tensor("strided_slice_5:0", shape=(), dtype=float64)
run(tf_t) :
 [[ 1.   2.   3.   4.   5.99]
  [ 2.   3.   4.   5.   6.99]
  [ 3.   4.   5.   6.   7.99]]
```

4）创建并输出三维张量：

```
a3dim = np.array([[[1,2],[3,4]],
                  [[5,6],[7,8]]
                 ])
print("a3dim Shape : ",a3dim.shape)

tf_t=tf.convert_to_tensor(a3dim,dtype=tf.float64)

print('tf_t : ',tf_t)
print('tf_t[0][0][0] : ',tf_t[0][0][0])
print('tf_t[1][1][1] : ',tf_t[1][1][1])
print('run(tf_t) : \n',tfs.run(tf_t))
```

输出结果如下：

```
a3dim Shape :  (2, 2, 2)
tf_t :  Tensor("Const_4:0", shape=(2, 2, 2), dtype=float64)
tf_t[0][0][0] :  Tensor("strided_slice_8:0", shape=(),
dtype=float64)
tf_t[1][1][1] :  Tensor("strided_slice_11:0", shape=(),
dtype=float64)
run(tf_t) :
 [[[ 1.  2.][ 3.  4.]]
  [[ 5.  6.][ 7.  8.]]]
```

> TensorFlow 可以将 NumPy 的 ndarray 无缝转换为 TensorFlow 张量，反之亦然。

1.2.7 变量

到目前为止，已经学习了如何创建各种张量对象：常量、操作和占位符。在使用 TensorFlow 创建和训练模型时，经常会遇到这种情况：需要将参数的值保存在内存中，以便程序运行时能更新，该内存位置由 TensorFlow 中的变量进行标识。

在 TensorFlow 中，变量是张量对象，包含在程序执行期间可被修改的值。

虽然 tf.Variable 与 tf.placeholder 类似，但两者之间存在细微差别，见表 1-3。

表 1-3

tf.placeholder	tf.Variable
tf.placeholder 定义的输入数据不会随时间变化	tf.Variable 定义的变量值会随时间推移而被改变
tf.placeholder 不需要在定义时设置初始值	tf.Variable 需要在定义时设定初始值

在 TensorFlow 中，可以使用 tf.Variable( ) 创建一个变量。下面来看一个基于占位符和变量的线性模型例子：

$$y=w\times x+b$$

1）将模型参数 w 和 b 定义为变量，初始值为 [.3] 和 [-0.3]：

```
w = tf.Variable([.3], tf.float32)
b = tf.Variable([-.3], tf.float32)
```

2）输入 x 被定义为占位符，输出 y 被定义为一个操作：

```
x = tf.placeholder(tf.float32)
y = w * x + b
```

3）输出 w、v、x 和 y：

```
print("w:",w)
print("x:",x)
print("b:",b)
print("y:",y)
```

得到如下输出：

```
w: <tf.Variable 'Variable:0' shape=(1,) dtype=float32_ref>
x: Tensor("Placeholder_2:0", dtype=float32)
b: <tf.Variable 'Variable_1:0' shape=(1,) dtype=float32_ref>
y: Tensor("add:0", dtype=float32)
```

输出显示 x 是占位符张量，y 是操作张量，而 w 和 b 分别是形状为 (1,) 和数据类型为 float32 的变量。

TensorFlow 会话在使用变量之前，必须对其进行初始化，可以通过运行其初始化操作来初始化单个变量。

例如，初始化变量 w：

```
tfs.run(w.initializer)
```

TensorFlow 提供了一个很方便的函数来初始化所有变量，在实践中经常使用该函数：

```
tfs.run(tf.global_variables_initializer())
```

也可以使用 tf.variables_initializer( ) 函数来初始化一组变量。

全局初始化函数也可以通过以下方式调用，这不会在会话对象的 run( ) 函数内调用：

```
tf.global_variables_initializer().run()
```

初始化变量后，运行模型，其输入为 x = [1, 2, 3, 4]：

```
print('run(y,{x:[1,2,3,4]}) : ',tfs.run(y,{x:[1,2,3,4]}))
```

得到以下输出结果：

```
run(y,{x:[1,2,3,4]}) :  [ 0.          0.30000001  0.60000002  0.90000004]
```

1.2.8 由库函数生成的张量

张量也可以由各种 TensorFlow 函数生成，这些生成的张量可以分配给常量或变量，也可以在初始化时提供给构造函数。

下面的代码生成了一个张量,其大小为100,值全为0,并输出该张量:

```
a=tf.zeros((100,))
print(tfs.run(a))
```

在定义张量时,TensorFlow 提供了不同类型的函数来填充张量:
- 使用相同的值填充所有元素;
- 使用序列填充元素;
- 使用概率分布函数填充元素,如正态分布函数或均匀分布函数。

1.2.8.1 使用相同的值填充张量元素

表 1-4 列出了一些张量生成库函数,它们能使用相同的值填充张量的所有元素。

表 1-4

张量生成函数	作用

zeros(
  shape,
  dtype=tf.float32,
  name=None
)
	按给定的形状创建张量,并将所有元素设置为零

zeros_like(
  tensor,
  dtype=None,
  name=None,
  optimize=True
)
	创建与参数形状相同的张量,并将所有元素设置为零

ones(
  shape,
  dtype=tf.float32,
  name=None
)
	按给定的形状创建张量,并将所有元素设置为 1

ones_like(
  tensor,
  dtype=None,
  name=None,
  optimize=True
)
	创建与参数形状相同的张量,并将所有元素设置为 1

fill(
  dims,
  value,
  name=None
)
	按给定的 dims 参数创建张量,并将所有元素的值设置为 value;例如,a = tf.fill([100], 0)

1.2.8.2 用序列填充张量元素

表 1-5 列出了一些张量生成函数,它们会用序列填充张量元素。

表 1-5

张量生成函数	作用
lin_space(　start, 　stop, 　num, 　name=None)	在 [start, stop] 范围内，所生成序列的元素个数为 num，用该序列生成一维张量。张量与 start 参数具有相同的数据类型 例如，a = tf.lin_space(1, 100, 10) 会生成值为 (1, 12, 23, 34, 45, 56, 67, 78, 89, 100) 的张量
range(　limit, 　delta=1, 　dtype=None, 　name='range')	在 [start，limit] 范围内，所生成序列的元素个数为 num，用该序列生成一维张量，delta 为增量。如果 dtype 参数没有被指定，那么张量与 start 参数的数据类型相同。这个函数有两个版本。在第二个版本中，如果 start 参数被省略，则该参数的值为 0。 例如，a = tf.range(1, 91, 10) 会生成值为 [1, 11, 21, 31, 41, 51, 61, 71, 81] 的张量。请注意这里的 limit 参数为 91，这意味着该值不包含在最终生成的序列中
range(　start, 　limit, 　delta=1, 　dtype=None, 　name='range')	

1.2.8.3 用随机分布填充张量元素

TensorFlow 能用随机分布函数的值来填充张量。

生成的分布受图的级别或操作级别种子的影响，图形种子使用 tf.set_random_seed 进行设置，而所有分布的操作种子则由参数 seed 给出。如果没有指定种子，会使用随机种子。

 有关 TensorFlow 中随机种子的更多详细信息，请访问以下链接：https://www.tensorflow.org/api_docs/python/tf/set_random_seed。

表 1-6 列出了一些能用随机值分布函数填充张量元素的张量生成函数。

表 1-6

张量生成函数	作用
random_normal(　shape, 　mean=0.0, 　stddev=1.0, 　dtype=tf.float32, 　seed=None, 　name=None)	生成指定形状的张量，用正态分布中的值填充：normal(mean, stddev)

（续）

张量生成函数	作用
truncated_normal(shape, mean=0.0, stddev=1.0, dtype=tf.float32, seed=None, name=None)	生成指定形状的张量，用截断正态分布的值填充：normal(mean, stddev)。截断意味着返回的值始终与平均值的距离小于两个标准偏差
random_uniform(shape, minval=0, maxval=None, dtype=tf.float32, seed=None, name=None)	生成指定形状的张量，用均匀分布的值填充：uniform([minval, maxval])
random_gamma(shape, alpha, beta=None, dtype=tf.float32, seed=None, name=None)	生成指定形状的张量，用伽马分布的值填充：gamma(alpha, beta) 关于 random_gamma 函数的更多细节可以在以下链接中找到：https://www.tensorflow.org/api_docs/python/ tf / random_gamma

1.2.9 通过 tf.get_variable( ) 获取变量

如果您定义的变量名称在之前已被定义过，则 TensorFlow 会引发异常。可使用 tf.get_variable( ) 函数代替 tf.Variable( )。如果变量存在，函数 tf.get_variable( ) 会返回现有的变量。如果变量不存在，会根据给定形状和初始值创建变量。例如：

```
w = tf.get_variable(name='w',shape=[1],dtype=tf.float32,initializer=[.3])
b = tf.get_variable(name='b',shape=[1],dtype=tf.float32,initializer=[-.3])
```

初始器（initializer）可以是上面例子中所示的张量或值的列表，或者是下面其中一种内置的初始器：

- tf.constant_initializer
- tf.random_normal_initializer
- tf.truncated_normal_initializer
- tf.random_uniform_initializer
- tf.uniform_unit_scaling_initializer
- tf.zeros_initializer
- tf.ones_initializer
- tf.orthogonal_initializer

在分布式 TensorFlow（代码能同时在多台计算机上运行）中，tf.get_variable( ) 能得到全局变量。为了得到局部变量，TensorFlow 有一个类似的函数：tf.get_local_variable( )。

共享或重用变量：获取已定义的变量有利于复用。但是，如果未使用 tf.variable_scope.reuse_variable( ) 或 tf.variable.scope(reuse = True) 设置复用标志，则会引发异常。

现在已经学会了如何定义张量、常量、操作、占位符和变量。下面将学习 TensorFlow 的另一个抽象层次，即把这些基本元素结合在一起形成基本的计算单元、数据流图或计算图。

1.3 数据流图或计算图

数据流图或计算图是 TensorFlow 的基本计算单位，从现在开始统称为计算图。计算图由节点和边组成，每个节点代表一个操作（tf.Operation），每条边代表在节点之间传递的张量（tf.Tensor）。

计算图是 TensorFlow 程序的基础。可创建带有节点的图，这些节点表示变量、常量、占位符和操作，并将其传递给 TensorFlow。TensorFlow 找到要触发或执行的第一个节点，这些节点的触发会导致其他节点被触发等。

因此，TensorFlow 程序由计算图上的两种操作组成：

- 构建计算图；
- 运行计算图。

TensorFlow 有默认的计算图，除非明确指定另一个计算图，否则会将新节点隐式添加到默认的计算图中。可以使用以下命令显式访问默认图形：

```
graph = tf.get_default_graph()
```

例如，如果想定义三个输入并将其相加以产生输出 $y=x_1+x_2+x_3$，可以使用如图 1-1 所示的计算图来表示。

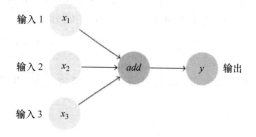

图 1-1

在 TensorFlow 中，图 1-1 相加操作对应的代码为 y = tf.add(x1 + x2 + x3)。

所创建的变量、常量和占位符会被添加到计算图中。然后创建一个会话对象来执行操作对象和张量对象。

下面将构建并执行一个计算图来计算 $y = w \times x + b$，就像在前面的例子中已经看到的那样：

```
# 假定线性模型 y = w * x + b
# 定义模型参数
w = tf.Variable([.3], tf.float32)
b = tf.Variable([-.3], tf.float32)
# 定义模型输入和输出
x = tf.placeholder(tf.float32)
y = w * x + b
output = 0
with tf.Session() as tfs:
    # 初始化并输出变量
    tf.global_variables_initializer().run()
    output = tfs.run(y,{x:[1,2,3,4]})
print('output : ',output)
```

在 with 块中创建和使用会话可确保会话在块完成时自动关闭。否则，会话必须使用明确的关闭命令 tfs.close( )，其中 tfs 是会话名称。

1.3.1 执行顺序和延迟加载

节点按照所依赖的顺序执行，如果节点 a 依赖于节点 b，那么当请求执行 b 时，将在 b 之前执行 a。除非节点本身或依赖于该节点的其他节点未被请求执行，否则节点不会执行。这也被称为延迟加载，即节点对象不被创建和初始化直到需要该节点。

有时可能想要控制图中节点的执行顺序，这可以通过 tf.Graph.control_dependencies( ) 函数来实现。例如，如果计算图具有节点 a、b、c 和 d，并且希望在 a 和 b 之前执行 c 和 d，则使用以下语句：

```
with graph_variable.control_dependencies([c,d]):
    # 这里的其他语句
```

这确保了前面 with 块的任何节点仅在节点 c 和 d 已被执行之后才会被执行。

1.3.2 跨计算设备执行计算图 -CPU 和 GPU

一个计算图可以分成多个部分，每部分可以放在不同的设备上（如 CPU 或 GPU）。可以使用以下命令列出可用于计算图执行的所有设备：

```
from tensorflow.python.client import device_lib
print(device_lib.list_local_devices())
```

得到以下输出（输出会有所不同，具体取决于系统的计算设备）：

```
[name: "/device:CPU:0"
device_type: "CPU"
memory_limit: 268435456
locality {
}
incarnation: 12900903776306102093
, name: "/device:GPU:0"
device_type: "GPU"
memory_limit: 611319808
locality {
  bus_id: 1
```

```
}
incarnation: 2202031001192109390
physical_device_desc: "device: 0, name: Quadro P5000, pci bus id:
0000:01:00.0, compute capability: 6.1"
]
```

关于上述输出需要注意的一点是：这里只显示了 1 个 CPU，而计算机有 8 个 CPU，原因是 TensorFlow 隐式地将代码分配到各个 CPU 上，因此默认情况下 CPU: 0 表示所有 CPU 都对 TensorFlow 有效。当 TensorFlow 开始执行计算图时，每个计算图会采用单独的线程来运行独立的路径，每个线程都在单独的 CPU 上运行。可以通过更改参数 inter_op_parallelism_threads 的值来限制用于此目的的线程数。同样，如果在独立路径内，操作能够在多个线程上运行，TensorFlow 将在多个线程上启动该特定操作。通过设置参数 intra_op_parallelism_threads 的值来更改线程数。

1.3.2.1 将计算图节点放置在特定的计算设备上

通过定义一个配置对象来放置变量日志，设置 log_device_placement 属性为 true，然后将此配置对象传递给会话，具体操作如下：

```
tf.reset_default_graph()

# 定义模型参数
w = tf.Variable([.3], tf.float32)
b = tf.Variable([-.3], tf.float32)
# 定义模型输入和输出
x = tf.placeholder(tf.float32)
y = w * x + b

config = tf.ConfigProto()
config.log_device_placement=True

with tf.Session(config=config) as tfs:
    # 初始化并输出变量 y
    tfs.run(global_variables_initializer())
    print('output',tfs.run(y,{x:[1,2,3,4]}))
```

在 Jupyter 笔记本控制台中获得以下输出：

```
b: (VariableV2): /job:localhost/replica:0/task:0/device:GPU:0
b/read: (Identity): /job:localhost/replica:0/task:0/device:GPU:0
b/Assign: (Assign): /job:localhost/replica:0/task:0/device:GPU:0
w: (VariableV2): /job:localhost/replica:0/task:0/device:GPU:0
w/read: (Identity): /job:localhost/replica:0/task:0/device:GPU:0
mul: (Mul): /job:localhost/replica:0/task:0/device:GPU:0
add: (Add): /job:localhost/replica:0/task:0/device:GPU:0
w/Assign: (Assign): /job:localhost/replica:0/task:0/device:GPU:0
init: (NoOp): /job:localhost/replica:0/task:0/device:GPU:0
x: (Placeholder): /job:localhost/replica:0/task:0/device:GPU:0
b/initial_value: (Const): /job:localhost/replica:0/task:0/device:GPU:0
Const_1: (Const): /job:localhost/replica:0/task:0/device:GPU:0
w/initial_value: (Const): /job:localhost/replica:0/task:0/device:GPU:0
Const: (Const): /job:localhost/replica:0/task:0/device:GPU:0
```

因此，默认情况下，TensorFlow 会在设备上创建能获得最高性能的变量和操作节点。通过使用 tf.device( ) 函数可以将变量和操作放置在指定的设备上。下面将计算图放在 CPU 上：

```
tf.reset_default_graph()

with tf.device('/device:CPU:0'):
    # 定义模型参数
    w = tf.get_variable(name='w',initializer=[.3], dtype=tf.float32)
    b = tf.get_variable(name='b',initializer=[-.3], dtype=tf.float32)
    # 定义输入和输出
    x = tf.placeholder(name='x',dtype=tf.float32)
    y = w * x + b

config = tf.ConfigProto()
config.log_device_placement=True

with tf.Session(config=config) as tfs:
    # 初始化并输出变量 y
    tfs.run(tf.global_variables_initializer())
    print('output',tfs.run(y,{x:[1,2,3,4]}))
```

在 Jupyter 控制台中，可以看到现在变量已放置在 CPU 上，执行过程也发生在 CPU 上：

```
b: (VariableV2): /job:localhost/replica:0/task:0/device:CPU:0
b/read: (Identity): /job:localhost/replica:0/task:0/device:CPU:0
b/Assign: (Assign): /job:localhost/replica:0/task:0/device:CPU:0
w: (VariableV2): /job:localhost/replica:0/task:0/device:CPU:0
w/read: (Identity): /job:localhost/replica:0/task:0/device:CPU:0
mul: (Mul): /job:localhost/replica:0/task:0/device:CPU:0
add: (Add): /job:localhost/replica:0/task:0/device:CPU:0
w/Assign: (Assign): /job:localhost/replica:0/task:0/device:CPU:0
init: (NoOp): /job:localhost/replica:0/task:0/device:CPU:0
x: (Placeholder): /job:localhost/replica:0/task:0/device:CPU:0
b/initial_value: (Const): /job:localhost/replica:0/task:0/device:CPU:0
Const_1: (Const): /job:localhost/replica:0/task:0/device:CPU:0
w/initial_value: (Const): /job:localhost/replica:0/task:0/device:CPU:0
Const: (Const): /job:localhost/replica:0/task:0/device:CPU:0
```

1.3.2.2 简单的放置

TensorFlow 遵循如下的简单规则（也称为简单放置规则）来将变量放置在设备上：

```
If the graph was previously run,
    then the node is left on the device where it was placed earlier
Else If the tf.device() block is used,
    then the node is placed on the specified device
Else If the GPU is present
    then the node is placed on the first available GPU
Else If the GPU is not present
    then the node is placed on the CPU
```

1.3.2.3 动态放置

也可以传递函数名称而不是设备字符串给 tf.device( )。在这种情况下，该函数必须返回

17

设备字符串，这一特性允许复杂算法将变量放置在不同设备上。例如，TensorFlow 可通过 tf.train.replica_device_setter( ) 函数来为设备提供循环设置器，稍后在下面讨论。

1.3.2.4　软放置

当在 GPU 上放置 TensorFlow 操作时，TF 必须具有该操作的 GPU 实现（称为内核）。如果内核不存在，则放置会导致运行时错误，另外，如果请求的 GPU 设备不存在，将会出现运行时错误。处理此类错误的最佳方法是：如果请求 GPU 设备导致错误，则允许将操作放置在 CPU 上。这可以通过设置以下 config 值来实现：

```
config.allow_soft_placement = True
```

1.3.2.5　GPU 内存处理

当开始运行 TensorFlow 会话时，默认情况下会获取所有 GPU 内存，即使只将操作和变量放置在多 GPU 系统中的一个 GPU 上。如果尝试同时运行另一个会话，则会出现内存不足的错误。这可以通过多种方式解决：

- 对于多 GPU 系统，请设置环境变量

```
CUDA_VISIBLE_DEVICES=<list of device idx>
os.environ['CUDA_VISIBLE_DEVICES']='0'
```

在此设置后执行的代码将仅能获取可见 GPU 的所有内存。
- 如果不希望会话获取 GPU 的所有内存，则可以使用配置选项 per_process_gpu_memory_fraction 来设置分配内存的百分比：

```
config.gpu_options.per_process_gpu_memory_fraction = 0.5
```

这将分配所有 GPU 设备的 50% 的内存。
- 也可以将上述两种策略结合起来，即只使用一个百分比，同时只让某些 GPU 可见。
- 还可以限制 TensorFlow 处理过程，以便在开始时仅获取所需的最小内存。随着进程进一步执行，为了允许内存增长，可以设置下面的配置选项。

```
config.gpu_options.allow_growth = True
```

此选项仅允许分配的内存增加，但内存不会释放回来。

在稍后的章节中，将学习在多个计算设备和多个节点上进行分布式计算。

1.3.3　多个计算图

可以创建自己的计算图，并在会话中执行。但是，建议不要创建和执行多个计算图，因为这样做有以下问题：

- 在同一个程序中创建和使用多个计算图需要多个 TensorFlow 会话，每个会话将消耗大量资源。
- 不能直接在计算图之间传递数据。

因此，推荐的方法是在一个计算图中包含多个计算子图。如果想使用自己的而不是默认的计算图，则可以使用 tf.graph( ) 命令来实现。这里给出创建自定义计算图 g 的一个例子，

并将其作为默认计算图来执行：

```
g = tf.Graph()
output = 0

# 假定线性模型 y = w * x + b

with g.as_default():
 # 定义模型参数
 w = tf.Variable([.3], tf.float32)
 b = tf.Variable([-.3], tf.float32)
 # 定义模型输入和输出
 x = tf.placeholder(tf.float32)
 y = w * x + b

with tf.Session(graph=g) as tfs:
 # 初始化并输出变量 y
 tf.global_variables_initializer().run()
 output = tfs.run(y,{x:[1,2,3,4]})

print('output : ',output)
```

1.4　TensorBoard

即使对于中等大小的问题，计算图的复杂度也会很高。代表复杂机器学习模型的大型计算图可能变得相当混乱且难以理解。可视化有助于轻松理解和解释计算图，从而加速 TensorFlow 程序的调试和优化。TensorFlow 带有一个能查看计算图的工具，该工具称为 TensorBoard。

TensorBoard 能可视化计算图的结构，提供统计分析并能绘制在执行计算图期间所捕获的摘要值。下面来看看在实践中它是如何工作的。

1.4.1　TensorBoard 最小的例子

1）首先为线性模型定义变量和占位符：

```
# 假定线性模型 y = w * x + b
# 定义模型参数
w = tf.Variable([.3], name='w',dtype=tf.float32)
b = tf.Variable([-.3], name='b', dtype=tf.float32)
# 定义模型输入和输出
x = tf.placeholder(name='x',dtype=tf.float32)
y = w * x + b
```

2）初始化会话，并在此会话的上下文中执行以下步骤：
- 初始化全局变量。
- 创建 tf.summary.FileWriter，该函数在包含默认计算图事件对应的 tflogs 文件夹中创建输出。
- 获取节点 y 的值，有效执行线性模型：

```
with tf.Session() as tfs:
    tfs.run(tf.global_variables_initializer())
    writer=tf.summary.FileWriter('tflogs',tfs.graph)
    print('run(y,{x:3}) : ', tfs.run(y,feed_dict={x:3}))
```

3）可得到如下结果：

```
run(y,{x:3}) :  [ 0.60000002]
```

在程序执行时，日志将被收集到 tflogs 文件夹中，以便 TensorBoard 进行可视化。打开命令行界面，转到 ch-01_TensorFlow_101 所在的文件夹，然后执行以下命令：

```
tensorboard --logdir='tflogs'
```

会看到类似下面的输出：

```
Starting TensorBoard b'47' at http://0.0.0.0:6006
```

打开浏览器并执行 http://0.0.0.0:6006。看到 TensorBoard 仪表盘后，不用理会所显示的任何错误或警告，只需单击顶部的图形选项卡即可。将看到如图 1-2 所示的结果。

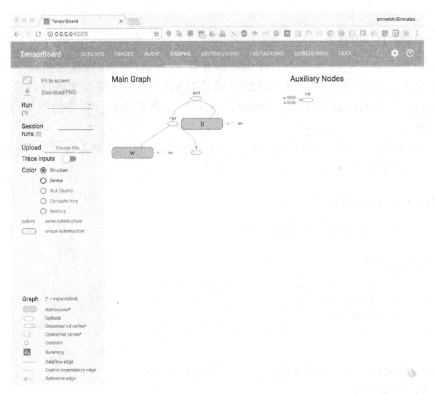

图 1-2

可以看到 TensorBoard 将第一个简单模型可视化为一个计算图，如图 1-3 所示。

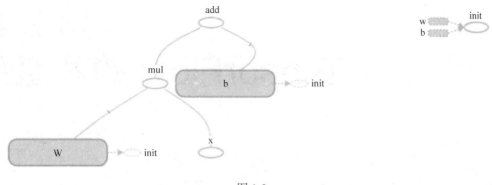

图 1-3

下面简单介绍 TensorBoard 的工作原理。

1.4.2 TensorBoard 的细节

TensorBoard 通过读取由 TensorFlow 生成的日志文件来工作。因此，需要修改此处定义的编程模型，以增加操作节点的额外信息，这些操作节点将在日志中生成 TensorBoard 可视化想要使用的信息。带有 TensorBoard 的编程模型或程序流程通常为：

1）创建通常的计算图。

2）创建节点摘要。通过 tf.summary 包生成的操作摘要作为节点的输出信息，这些信息是人们希望收集和分析的。

3）在运行模型节点时就会得到节点摘要信息，通常，可以使用函数 tf.summary.merge_all( ) 来合并所有节点的汇总信息，并汇总成一个节点摘要。执行这个合并的节点就基本上执行所有的汇总节点。合并的汇总节点会生成一个经过序列化后的 Summary ProtocolBuffers 对象，该对象包含了所有摘要。

4）通过将 Summary ProtocolBuffers 对象传递给 tf.summary.FileWriter 对象以便将事件日志写入磁盘。

5）启动 TensorBoard 来分析需要可视化的数据。

在本节中，没有创建节点的摘要，只是以非常简单的方式使用了 TensorBoard。本书稍后会介绍 TensorBoard 的高级用法。

1.5 总结

本章简要介绍了 TensorFlow 库，并介绍了可用于构建 TensorFlow 计算图的 TensorFlow 数据模型元素，例如常量、变量和占位符。也介绍了如何从 Python 对象创建张量，张量对象也可以通过 TensorFlow 中各种库函数的特定值、序列或随机值分布来生成。

TensorFlow 编程模型由构建和执行计算图组成，计算图有节点和边。节点表示操作，边表示张量，这些张量会将数据从一个节点传递给到另一个节点。本章还介绍了如何创建和执行计算图，执行顺序以及如何在不同的计算设备（如 GPU 和 CPU）上执行计算图。本章最后介绍了用于可视化 TensorFlow 计算图的工具 TensorBoard。

下一章将探讨一些构建在 TensorFlow 之上的高级库，并将介绍如何快速构建模型。

第 2 章 TensorFlow 的高级库

TensorFlow 有几个高级库和接口（API），可以允许用户轻松地构建模型并用少量代码训练模型，这些高级库和接口包括 TF Learn、TF Slim、Sonnet、PrettyTensor、Keras 和最近发布的 TensorFlow Estimator 等。

本章将介绍以下高级库，并在下一章专门介绍 Keras：

- TF Estimator；
- TF Slim；
- TF Learn；
- PrettyTensor；
- Sonnet。

本章将使用这 5 个库来构建 MNIST 数据集模型的示例。请读者不要担心无法理解这些模型的细节，因为从第 4 章开始会介绍相关的内容。

本章的代码示例包含在代码库的 Jupyter 笔记本 ch-02_TF_High_Level_Libraries。读者可尝试修改这些代码来进行试验。

2.1 TF Estimator

TF Estimator（其前身是 TFLearn）是一个高级 API，它封装了训练、评估、预测和导出功能，从而可以轻松创建和训练模型。最近 TensorFlow 以 TF Estimator 的新名称重新命名并发布了 TFLearn 包，这可能是为了避免与来自 tflearn.org 的 TFLearn 包混淆。TF Estimator API 对最初的 TF Learn 包进行了重大改进，这些包在 KDD 17 的会议论文中进行了描述，论文可在以下链接中找到：https：//doi.org/10.1145/3097983.3098171。

TF Estimator 接口设计受到了流行的机器学习库 SciKit Learn 的启发，允许从各种可用模型创建 Estimator 对象，而任何 Estimator 上都有 4 个主要功能：

- estimator.fit( )
- estimator.evaluate( )
- estimator.predict( )
- estimator.export( )

这些功能的名称是不言自明的，Estimator 对象表示模型，但模型本身是从提供给 Estimator 的模型定义函数创建的。

可以在图 2-1 中描绘 Estimator 对象及其接口。

使用 Estimator API 而不是 TensorFlow 内核的所有内容，就无需担心计算图、会话、初始化变量或其他低级细节。在撰写本书时，TensorFlow 提供了以下预建（pre-built）的 Estimator：

- tf.contrib.learn.KMeansClustering
- tf.contrib.learn.DNNClassifier
- tf.contrib.learn.DNNRegressor
- tf.contrib.learn.DNNLinearCombinedRegressor
- tf.contrib.learn.DNNLinearCombinedClassifier
- tf.contrib.learn.LinearClassifier
- tf.contrib.learn.LinearRegressor
- tf.contrib.learn.LogisticRegressor

TF Estimator API 中的简单工作流程为：

1）找到与问题相关的预建 Estimator；
2）编写导入数据集的函数；
3）定义数据中包含特征的列；
4）创建在步骤 1 中选择的 Estimator 的实例；
5）训练 Estimator；
6）使用训练过的 Estimator 进行评估或预测。

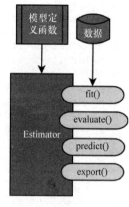

图 2-1

  在下一章讨论的 Keras 库提供了一个方便的功能，可以将 Keras 模型转换为 Estimator：keras.estimator.model_to_estimator( )。

在 ch-02_TF_High_Level_Libraries 中提供了 MNIST 分类示例的完整代码。TF Estimator 在 MNIST 示例的输出结果如下：

```
INFO:tensorflow:Using default config.
WARNING:tensorflow:Using temporary folder as model directory: /tmp/tmprvcqgu07
INFO:tensorflow:Using config: {'_save_checkpoints_steps': None, '_task_type': 'worker', '_save_checkpoints_secs': 600, '_service': None, '_task_id': 0, '_master': '', '_session_config': None, '_num_worker_replicas': 1, '_keep_checkpoint_max': 5, '_cluster_spec': <tensorflow.python.training.server_lib.ClusterSpec object at 0x7ff9d15f5fd0>, '_keep_checkpoint_every_n_hours': 10000, '_log_step_count_steps': 100, '_is_chief': True, '_save_summary_steps': 100, '_model_dir': '/tmp/tmprvcqgu07', '_num_ps_replicas': 0, '_tf_random_seed': None}
INFO:tensorflow:Create CheckpointSaverHook.
INFO:tensorflow:Saving checkpoints for 1 into /tmp/tmprvcqgu07/model.ckpt.
INFO:tensorflow:loss = 2.4365, step = 1
INFO:tensorflow:global_step/sec: 597.996
INFO:tensorflow:loss = 1.47152, step = 101 (0.168 sec)
INFO:tensorflow:global_step/sec: 553.29
```

```
INFO:tensorflow:loss = 0.728581, step = 201 (0.182 sec)
INFO:tensorflow:global_step/sec: 519.498
INFO:tensorflow:loss = 0.89795, step = 301 (0.193 sec)
INFO:tensorflow:global_step/sec: 503.414
INFO:tensorflow:loss = 0.743328, step = 401 (0.202 sec)
INFO:tensorflow:global_step/sec: 539.251
INFO:tensorflow:loss = 0.413222, step = 501 (0.181 sec)
INFO:tensorflow:global_step/sec: 572.327
INFO:tensorflow:loss = 0.416304, step = 601 (0.174 sec)
INFO:tensorflow:global_step/sec: 543.99
INFO:tensorflow:loss = 0.459793, step = 701 (0.184 sec)
INFO:tensorflow:global_step/sec: 687.748
INFO:tensorflow:loss = 0.501756, step = 801 (0.146 sec)
INFO:tensorflow:global_step/sec: 654.217
INFO:tensorflow:loss = 0.666772, step = 901 (0.153 sec)
INFO:tensorflow:Saving checkpoints for 1000 into /tmp/tmprvcqgu07/model.ckpt.
INFO:tensorflow:Loss for final step: 0.426257.
INFO:tensorflow:Starting evaluation at 2017-12-15-02:27:45
INFO:tensorflow:Restoring parameters from /tmp/tmprvcqgu07/model.ckpt-1000
INFO:tensorflow:Finished evaluation at 2017-12-15-02:27:45
INFO:tensorflow:Saving dict for global step 1000: accuracy = 0.8856, global_step = 1000, loss = 0.40996

{'accuracy': 0.88559997, 'global_step': 1000, 'loss': 0.40995964}
```

在第 5 章中将看到如何使用 TensorFlow 内核函数来创建此类模型。

2.2 TF Slim

TF Slim 是一个构建在 TensorFlow 内核之上的轻量级库,它可用于定义和训练模型。TF Slim 可以与其他 TensorFlow 低级库和高级库(如 TF Learn)结合使用。TF Slim 是 TensorFlow 的一部分:tf.contrib.slim。运行以下命令来检查 TF Slim 的安装工作是否正常:

```
python3 -c 'import tensorflow.contrib.slim as slim; eval = slim.evaluation.evaluate_once'
```

TF Slim 提供了几个模块,可以独立安装和使用它们,并且也可与其他 TensorFlow 软件包混合使用。例如,在撰写本书时,TF Slim 的主要模块如表 2-1 所示。

表 2-1

TF Slim 模块	模块说明
arg_scope	提供了一种将元素应用于在作用域下所定义的所有计算图节点的机制
层	提供几种不同类型的神经网络层,例如 fully_connected、conv2d 等
损失函数	提供训练优化器的损失函数
学习	提供训练模型的函数
评测	提供评估函数
度量	提供用于评估模型的度量函数
正则化	提供创建正则化方法的函数
变量	为创建变量提供函数
网络模型	提供各种预先构建和预先训练的模型,如 VGG16、InceptionV3、ResNet

TF Slim 中的简单工作流程如下：
1）使用 slim 层创建模型。
2）为创建的层提供输入数据以实例化模型。
3）使用 logits 和标签来定义损失函数。
4）使用函数 get_total_loss( ) 来获取总损失函数。
5）创建优化器。
6）使用 slim.learning.create_train_op( )、total_loss 和 optimizer 来创建训练函数。
7）使用前面定义的函数 slim.learning.train( ) 和训练函数运行训练。

在 ch-02_TF_High_Level_Libraries 中提供了 MNIST 分类示例的完整代码。TF Slim MNIST 的输出示例如下：

```
INFO:tensorflow:Starting Session.
INFO:tensorflow:Saving checkpoint to path ./slim_logs/model.ckpt
INFO:tensorflow:global_step/sec: 0
INFO:tensorflow:Starting Queues.
INFO:tensorflow:global step 100: loss = 2.2669 (0.010 sec/step)
INFO:tensorflow:global step 200: loss = 2.2025 (0.010 sec/step)
INFO:tensorflow:global step 300: loss = 2.1257 (0.010 sec/step)
INFO:tensorflow:global step 400: loss = 2.0419 (0.009 sec/step)
INFO:tensorflow:global step 500: loss = 1.9532 (0.009 sec/step)
INFO:tensorflow:global step 600: loss = 1.8733 (0.010 sec/step)
INFO:tensorflow:global step 700: loss = 1.8002 (0.010 sec/step)
INFO:tensorflow:global step 800: loss = 1.7273 (0.010 sec/step)
INFO:tensorflow:global step 900: loss = 1.6688 (0.010 sec/step)
INFO:tensorflow:global step 1000: loss = 1.6132 (0.010 sec/step)
INFO:tensorflow:Stopping Training.
INFO:tensorflow:Finished training! Saving model to disk.
final loss=1.6131552457809448
```

正如从输出中看到的，函数 slim.learning.train( ) 会将训练输出保存在指定日志目录中的检查点文件中。如果重新训练，会首先检查检查点是否存在，默认将从检查点文件恢复训练。

在撰写本书时，以下链接中的 TF Slim 文档页面为空：https://www.tensorflow.org/api_docs/python/tf/contrib/slim。但是，有些文档可以在以下链接的源代码中找到：https://github.com/tensorflow/tensorflow/tree/r1.4/tensorflow/contrib/slim。

在后面章节中将会使用 TF Slim 来学习如何使用预先训练的模型（如 VGG16 和 Inception V3）。

2.3　TFLearn

TFLearn 是 Python 中的一个模块库，构建于 TensorFlow 内核之上。

> TFLearn 不同于 TensorFlow Learn 包，TensorFlow Learn 包也被称为 TF Learn（TF 和 Learn 之间有一个空格）。TFLearn 可在以下链接中找到：http://tflearn.org；源代码可在 GitHub 找到：https://github.com/tflearn/tflearn。

使用以下命令可以在 Python 3 中安装 TFLearn：

```
pip3 install tflearn
```

> 要在其他环境或源代码中安装 TFLearn，请参阅以下链接：http://tflearn.org/installation/。

TFLearn 的简单工作流程如下：

1）首先创建输入层。
2）传递输入对象以创建更多的层。
3）添加输出层。
4）使用 Estimator 层（regression）创建网络。
5）从前面创建的网络创建模型。
6）使用 model.fit( ) 方法训练模型。
7）使用训练好的模型来预测或评估。

2.3.1 创建 TFLearn 层

下面来学习如何用 TFLearn 创建神经网络模型的层。

1）首先创建输入层：

```
input_layer = tflearn.input_data(shape=[None,num_inputs]
```

2）传递输入对象以创建更多网络层：

```
layer1 = tflearn.fully_connected(input_layer,10,
                                 activation='relu')
layer2 = tflearn.fully_connected(layer1,10,
                                 activation='relu')
```

3）添加输出层：

```
output = tflearn.fully_connected(layer2,n_classes,
                                 activation='softmax')
```

4）使用 Estimator 层（regression）创建最终网络：

```
net = tflearn.regression(output,
                         optimizer='adam',
                         metric=tflearn.metrics.Accuracy(),
                         loss='categorical_crossentropy'
                         )
```

下面介绍 TFLearn 提供的几种网络层。

2.3.1.1 TFLearn 内核层

TFLearn 在 tflearn.layers.core 模块中提供的层见表 2-2。

表 2-2

层 类 型	描 述
input_data	该层用于指定神经网络的输入层
fully_connected	该层用来定义一个层,其中所有神经元都连接到前一层中的所有神经元
dropout	该层用于指定 dropout 正则化。输入元素按 1 / keep_prob 进行丢弃,并保持预期总和不变
custom_layer	该层用于指定要应用于输入的定制函数。它封装了定制函数,并将该函数作为一个层来展现
reshape	该层将输入重新映射成给定形状的输出
flatten	该层将输入张量转换为二维张量
activation	该层将指定应用于输入张量的激活函数
single_unit	该层将线性函数应用于输入
highway	该层实现完全连接的 highway 函数
one_hot_encoding	该层将数字标签转换为相应的二进制向量独热 (one-hot) 编码表示
time_distributed	该层将指定的函数应用于输入张量的每个时间步
multi_target_data	该层创建并连接多个占位符。具体而言,当使用多源(multiple sources)目标时,会使用到该层

2.3.1.2 TFLearn 卷积层

TFLearn 在 tflearn.layers.conv 模块中提供的层见表 2-3。

表 2-3

层 类 型	描 述
conv_1d	该层将一维卷积作为输入数据
conv_2d	该层将二维卷积作为输入数据
conv_3d	该层将三维卷积作为输入数据
conv_2d_transpose	该层将 conv2_d 的转置作为输入数据
conv_3d_transpose	该层将 conv3_d 的转置作为输入数据
atrous_conv_2d	该层计算二维 atrous 卷积
grouped_conv_2d	该层计算深度 (depth-wise) 二维卷积
max_pool_1d	该层计算一维最大池化
max_pool_2d	该层计算二维最大池化
avg_pool_1d	该层计算一维平均池化
avg_pool_2d	该层计算二维平均池化
upsample_2d	该层逐行和逐列地应用于二维重复操作
upscore_layer	该层实现了 http://arxiv.org/abs/1411.4038 中指定的 upscore
global_max_pool	该层实现全局最大池化操作

(续)

层 类 型	描 述
global_avg_pool	该层实现全局平均池化操作
residual_block	该层实现用残差块来创建深度残差网络
residual_bottleneck	该层实现了深度残差网络的残差瓶颈块
resnext_block	该层实现 Resnext 块

2.3.1.3 TFLearn 循环 (recurrent) 层

TFLearn 在 tflearn.layers.recurrent 模块中提供的层见表 2-4。

表 2-4

层 类 型	描 述
simple_rnn	该层实现了简单的循环神经网络（RNN）模型
bidirectional_rnn	该层实现了双向 RNN 模型
lstm	该层实现了 LSTM 模型
gru	该层实现 GRU 模型

2.3.1.4 TFLearn 归一化层

TFLearn 在 tflearn.layers.normalization 模块中提供的层见表 2-5。

表 2-5

层 类 型	描 述
batch_normalization	对每一批次，该层会归一化前一层由激活函数得到的输出结果
local_response_normalization	该层实现 LR 归一化
l2_normalization	该层对输入张量进行 L2 归一化

2.3.1.5 TFLearn 嵌入层

TFLearn 在 tflearn.layers.embedding_ops 模块中只提供一个层，见表 2-6。

表 2-6

层 类 型	描 述
embedding	该层为整数或浮点 ID 序列实现嵌入函数

2.3.1.6 TFLearn 合并层

TFLearn 在 tflearn.layers.merge_ops 模块中提供的层见表 2-7。

表 2-7

层 类 型	描 述
merge_outputs	该层将一组张量合并为单一张量，通常用于合并相同形状的输出张量
merge	该层将一组张量合并为单一张量；可以指定合并需要的轴

2.3.1.7 TFLear 的 estimator 层

TFLearn 在 tflearn.layers.estimator 模块中只提供一个层，见表 2-8。

表 2-8

层 类 型	描 述
regression	该层实现线性或逻辑回归

在创建回归层时，可以指定优化器、损失函数、度量函数。

TFLearn 在 tflearn.optimizers 包中提供以下类型的优化器函数：

- SGD
- RMSprop
- Adam
- Momentum
- AdaGrad
- Ftrl
- AdaDelta
- ProximalAdaGrad
- Nesterov

可以通过扩展 tflearn.optimizers.Optimizer 基类来自定义优化器。

TFLearn 在 tflearn.metrics 模块中提供以下度量函数：

- Accuracy 或 accuracy_op
- Top_k 或 top_k_op
- R2 或 r2_op
- WeightedR2 或 weighted_r2_op
- binary_accuracy_op

可以通过扩展 tflearn.metrics.Metric 基类来自定义度量函数。

TFLearn 在 tflearn.objectives 模块中提供以下损失函数（也称为目标函数）：

- softymax_categorical_crossentropy
- categorical_crossentropy
- binary_crossentropy
- weighted_crossentropy
- mean_square
- hinge_loss

- roc_auc_score
- weak_cross_entropy_2d

在创建输入层、隐藏层和输出层时，可以指定要应用于输出的激活函数。TFLearn 在 tflearn.activations 模块中提供以下激活函数：

- linear
- tanh
- sigmoid
- softmax
- softplus
- softsign
- relu
- relu6
- leaky_relu
- prelu
- elu
- crelu
- selu

2.3.2 创建 TFLearn 模型

利用前面创建的网络来创建模型（创建 TFLearn 层的步骤 4）：

```
model = tflearn.DNN(net)
```

2.3.2.1 TFLearn 模型的类型

TFLearn 提供两种不同类型的模型：
- DNN（深度神经网络）模型：该类通过神经网络层来创建多层感知器网络。
- SequenceGenerator 模型：该类允许创建一个深度神经网络，该网络可以生成序列。

2.3.3 训练 TFLearn 模型

创建完模型后，可使用 model.fit( ) 方法来训练模型：

```
model.fit(X_train,
          Y_train,
          n_epoch=n_epochs,
          batch_size=batch_size,
          show_metric=True,
          run_id='dense_model')
```

2.3.4 使用 TFLearn 模型

使用训练好的模型来进行预测或评估：

```
score = model.evaluate(X_test, Y_test)
print('Test accuracy:', score[0])
```

在 ch-02_TF_High_Level_Libraries 中提供了 TFLearn MNIST 分类示例的完整代码。TFLearn MNIST 的示例输出如下：

```
Training Step: 5499  | total loss: 0.42119 | time: 1.817s
| Adam | epoch: 010 | loss: 0.42119 - acc: 0.8860 -- iter: 54900/55000
Training Step: 5500  | total loss: 0.40881 | time: 1.820s
| Adam | epoch: 010 | loss: 0.40881 - acc: 0.8854 -- iter: 55000/55000
--
Test accuracy: 0.9029
```

可以从以下链接获得更多关于 TFLearn 的信息：http：//tflearn.org/。

2.4 PrettyTensor

PrettyTensor 在 TensorFlow 内核上进行了轻量级封装。PrettyTensor 可通过链式语法（chainable syntax）来定义神经网络。例如，可以通过链接层来创建模型，具体创建如以下代码所示：

```
model = (X.
         flatten().
         fully_connected(10).
         softmax_classifier(n_classes, labels=Y))
```

在 Python 3 中，可使用以下命令来安装 PrettyTensor：

```
pip3 install prettytensor
```

PrettyTensor 通过 apply( ) 方法提供了一个非常轻量级且可扩展的接口。apply(function, arguments) 方法可将任何函数都链接到 PrettyTensor 对象中。PrettyTensor 将调用该函数，并提供当前张量来作为该函数的第一个参数。

可以使用修饰器（decorator）@ prettytensor.register 来添加用户创建的函数。具体细节可以在 https://github.com/google/prettytensor 找到。

在 PrettyTensor 中定义和训练模型的工作流程如下：
1）获取数据。
2）定义超参数和参数。
3）定义输入和输出。
4）定义模型。
5）定义评估器、优化器和训练器函数。

6）创建运行器对象。

7）在 TensorFlow 会话中，使用 runner.train_model( ) 方法来训练模型。

8）在同一个会话中，使用 runner.evaluate_model( ) 方法来评估模型。

PrettyTensor MNIST 分类示例的完整代码在 ch-02_TF_High_Level_Libraries 中可找到。

PrettyTensor MNIST 的输出结果如下：

```
[1] [2.5561881]
[600] [0.3553167]
Accuracy after 1 epochs 0.8799999952316284

[601] [0.47775066]
[1200] [0.34739292]
Accuracy after 2 epochs 0.8999999761581421

[1201] [0.19110668]
[1800] [0.17418651]
Accuracy after 3 epochs 0.8999999761581421

[1801] [0.27229539]
[2400] [0.34908807]
Accuracy after 4 epochs 0.8700000047683716

[2401] [0.40000191]
[3000] [0.30816519]
Accuracy after 5 epochs 0.8999999761581421

[3001] [0.29905257]
[3600] [0.41590339]
Accuracy after 6 epochs 0.8899999856948853

[3601] [0.32594997]
[4200] [0.36930788]
Accuracy after 7 epochs 0.8899999856948853

[4201] [0.26780865]
[4800] [0.2911002]
Accuracy after 8 epochs 0.8899999856948853

[4801] [0.36304188]
[5400] [0.39880857]
Accuracy after 9 epochs 0.8999999761581421

[5401] [0.1339224]
[6000] [0.14993289]
Accuracy after 10 epochs 0.8899999856948853
```

2.5 Sonnet

Sonnet 是一个用 Python 编写的面向对象库，于 2017 年由 DeepMind 发布。

Sonnet 将下面两种通过对象创建计算图的方式彻底分离开：

- 对象配置调用模块。
- 对象与计算图的连接。

使用以下命令可以在 Python 3 中安装 Sonnet：

```
pip3 install dm-sonnet
```

 通过以下链接可以从源代码安装 Sonnet：
https://github.com/deepmind/sonnet/blob/master/docs/INSTALL.md。

这些模块被定义为抽象类 sonnet.AbstractModule 的子类。在撰写本书时，Sonnet 提供的模块见表 2-9。

表 2-9

基本模块	AddBias、BatchApply、BatchFlatten、BatchReshape、FlattenTrailingDimensions、Linear、MergeDims、SelectInput、SliceByDim、TileByDim 和 TrainableVariable
循环模块	DeepRNN、ModelRNN、VanillaRNN、BatchNormLSTM、GRU 和 LSTM
循环 + ConvNet 模块	Conv1DLSTM 和 Conv2DLSTM
ConvNet 模块	Conv1D、Conv2D、Conv3D、Conv1DTranspose、Conv2DTranspose、Conv3DTranspose、DepthWiseConv2D、InPlaneConv2D 和 SeparableConv2D
ResidualNets	Residual、ResidualCore 和 SkipConnectionCore
其他	BatchNorm、LayerNorm、clip_gradient 和 scale_gradient

可以通过创建一个 sonnet.AbstractModule 的子类来自定义新模块，另一种方法是通过函数创建模块（不推荐这种方法），即通过将函数传递给封装的模块来创建一个 sonnet.Module 对象。

通过 Sonnet 库建立模型的工作流程如下：

1）为继承于 sonnet.AbstractModule 的数据集和网络体系结构创建类。在这个例子中，会创建一个 MNIST 类和一个 MLP 类。

2）定义参数和超参数。

3）根据前面定义的数据集类来定义测试和训练数据集。

4）使用定义的网络类来定义模型，比如 model = MLP([20, n_classes]) 会创建一个 MLP 网络，其包含两层，分别包括 20 个和 n_classes 个神经元。

5）为训练数据集和测试数据集定义 y_hat 占位符。

6）定义训练数据集和测试数据集的损失占位符。

7）使用训练数据集损失占位符来定义优化器。

8）在 TensorFlow 会话中按给定的步数执行损失函数以优化参数。

在 ch-02_TF_High_Level_Libraries 中提供了 Sonnet MNIST 分类示例的完整代码。每个类的 _init_ 方法会初始化类以及相关的超级类。_build 方法创建并返回数据集或模型对象。Sonnet MNIST 示例的输出结果如下：

```
Epoch : 0 Training Loss : 236.79913330078125
Epoch : 1 Training Loss : 227.3693084716797
Epoch : 2 Training Loss : 221.96337890625
Epoch : 3 Training Loss : 220.99142456054688
Epoch : 4 Training Loss : 215.5921173095703
Epoch : 5 Training Loss : 213.88958740234375
Epoch : 6 Training Loss : 203.7091064453125
Epoch : 7 Training Loss : 204.57427978515625
Epoch : 8 Training Loss : 196.17218017578125
Epoch : 9 Training Loss : 192.3954315185547
Test loss : 192.8847198486328
```

由于神经网络中计算的随机性，输出的结果可能会有所不同。这些内容是对 Sonnet 模块的简单介绍。

> 有关 Sonnet 的更多详细信息，可以浏览以下链接：
> https://deepmind.github.io/sonnet/.

2.6 总结

本章学习了构建在 TensorFlow 上的一些高级库，了解了 TF Estimator、TF Slim、TFLearn、PrettyTensor 和 Sonnet，并基于这 5 个高级库分别实现了 MNIST 分类。如果无法理解模型的细节，请不要担心，因为在下一章还会介绍基于 MNIST 所构建的模型。

表 2-10 总结了本章提供的库和框架。

表 2-10

高级库	文档链接	源代码链接	pip3 安装包
TF Estimator	https://www.tensorflow.org/get_started/estimator	https://github.com/tensorflow/tensorflow/tree/master/tensorflow/python/estimator	TensorFlow 预装库
TF Slim	https://github.com/tensorflow/tensorflow/tree/r1.4/tensorflow/contrib/slim	https://github.com/tensorflow/tensorflow/tree/r1.4/tensorflow/contrib/slim	TensorFlow 预装库
TFLearn	http://tflearn.org/	https://github.com/tflearn/tflearn	tflearn
PrettyTensor	https://github.com/google/prettytensor/tree/master/docs	https://github.com/google/prettytensor	prettytensor
Sonnet	https://deepmind.github.io/sonnet/	https://github.com/deepmind/sonnet	dm-sonnet

下一章将介绍 Keras，这是用于创建和训练 TensorFlow 模型的高级库，它非常受欢迎。

第 3 章
Keras 101

Keras 是一个高级的深度学习库，该库将 TensorFlow 作为后端。TensorFlow 团队已将 Keras 作为模块 tf.keras 包含在 TensorFlow 的内核中。在撰写本书时，Keras 除了 TensorFlow 之外，也支持 Theano 和 CNTK。

Keras 的以下指导原则使它在深度学习社区中非常受欢迎：
- 极简主义，提供一致和简单的 API；
- 模块化，允许各种元素的表示作为可拔插模块；
- 可扩展性，以类和函数的形式添加新模块；
- 代码和模型配置都是基于本地化 Python（Python-native）；
- 支持 CNN、RNN 或两者结合的开箱（Out-of-the-box）通用的网络体系结构。

本书后面部分将学习如何使用低级 TensorFlow API 和高级 Keras API 来构建不同类型的深度学习模型和机器学习模型。

本章将介绍以下主题：
- 安装 Keras；
- 在 Keras 中创建模型的过程；
- 使用序列化和功能性 API 创建 Keras 模型；
- Keras 的层；
- 使用序列化和功能性 API 创建和添加层；
- 编译 Keras 模型；
- 训练 Keras 模型；
- 使用 Keras 模型；
- 使用 Keras 模型进行预测；
- 基于 MNIST 数据集的 Keras 顺序模型示例。

3.1 安装 Keras

在 Python 3 中，可以使用以下命令安装 Keras：

```
pip3 install keras
```

若要在其他环境或来源安装Keras,请参阅以下链接:https://keras.io/#installation。

3.2 Keras的神经网络模型

Keras的神经网络模型被定义为层的图(graph)。Keras中的模型可以使用序列化API或功能性API来创建;功能性API和序列化API都可用于构建任何类型的模型。功能性API能构建具有多个输入、多个输出和共享层,这使得构建复杂模型变得更加容易。

因此,通常可以看到工程师们使用序列化API来构建简单模型,这些简单模型由简单的层构建而成,功能性API用于构建分支和图共享的复杂模型。笔者还观察到,使用功能性API构建简单模型可以更轻松地将模型扩展为具有分支和共享的复杂模型。因此,对于笔者的工作而言,总是使用功能性API。

3.2.1 在Keras中创建模型的过程

Keras的简单工作流程如下:

1)创建模型;
2)创建模型并添加层;
3)编译模型;
4)训练模型;
5)使用该模型进行预测或评估。

下面来介绍每个步骤。

可以在Jupyter笔记本ch-03_Keras_101中看到本章的代码示例。为了研究各种功能,可尝试修改这些代码。

3.3 创建Keras模型

Keras模型可以使用序列化API或功能性API来创建。下面给出以两种方式创建模型的示例。

3.3.1 用于创建Keras模型的序列化API

在序列化API中,使用以下代码可创建空模型:

```
model = Sequential()
```

现在可以将这些层添加到此模型中,将会在下一部分使用这些层。或者,也可以将所有层作为列表传递给构造函数。下面这个例子将层传递给构造函数来添加4个层:

```
model = Sequential([ Dense(10, input_shape=(256,)),
                     Activation('tanh'),
                     Dense(10),
                     Activation('softmax')
                   ])
```

3.3.2 用于创建 Keras 模型的功能性 API

在功能性 API 中，将模型创建为 Model 类的一个实例，该实例接受输入和输出参数。输入和输出参数分别表示一个或多个输入和输出张量。

通过功能性 API 来实例化模型的代码如下：

```
model = Model(inputs=tensor1, outputs=tensor2)
```

在上面的代码中，tensor1 和 tensor2 是张量或可以像张量那样进行处理的对象，例如 Keras 中网络层对象。

如果有多个输入和输出张量，则可以作为列表传递，如下所示：

```
model = Model(inputs=[i1,i2,i3], outputs=[o1,o2,o3])
```

3.4 Keras 的层

Keras 提供了几种内置的网络层，以便于构建网络架构。下面是在编写本书时对 Keras 2 提供的各种网络层的总结和描述。

3.4.1 Keras 内核层

Keras 内核层实现的基本操作几乎会在所有的网络架构中使用。表 3-1 对 Keras 2 提供的网络层进行了总结。

表 3-1

网络层名称	描述
Dense	这是一个全连接的简单神经网络层。该层通过以下函数作为输出：*activation((inputs x weights)+bias)*，其中 *activation* 是传递给神经网络层的激活函数，默认为 None
Activation	该层将指定用于输出的激活函数。该层产生以下输出：*activation(inputs)*，其中 *activation* 是指传递给网络层的激活函数 以下激活函数可用于该网络层：softmax、elu、selu、softplus、softsign、relu、tanh、sigmoid、hard_sigmoid 和 linear
Dropout	该层对输入按给定的比率进行 dropout 正则化
Flatten	该层使输入变平，比如将三维的输入变平就会产生一维输出
Reshape	该层将输入转换为指定的形状
Permute	该层按照指定模式对输入大小进行重新排序
RepeatVector	该层按给定的次数重复输入。因此，如果输入是形状为（#samples, #features）的二维张量，并且该层被重复 *n* 次，则输出的形状为三维张量（#samples, n, #features）
Lambda	该层将提供的功能作为图层进行包装。因此，输入将通过提供的自定义函数传递以产生输出。Keras 用户可以利用该层将自定义功能添加为图层，从而使 Keras 具有终极扩展性

（续）

网络层名称	描述
ActivityRegularization	该层适用于L1正则化、L2正则化，或两者相结合的正则化。该层作为激活层的输出或具有激活函数的神经网络层的输出
Masking	该层会跳过输入张量的所有时间步（time step），其中输入张量中的所有值等于掩码值，神经网络层提供参数可让用户指定掩码值

3.4.2 Keras 卷积层

这些层实现 CNN 中不同类型的卷积、采样和裁剪操作，见表 3-2。

表 3-2

网络层名称	描述
Conv1D	该层将单个空间或时间维度上的卷积应用于输入
Conv2D	该层对输入进行二维卷积
SeparableConv2D	该层在每个输入通道上应用深度方向的空间卷积，然后逐点卷积，将所得输出通道混合在一起
Conv2DTranspose	该层将卷积的形状恢复为产生这些卷积的输入的形状
Conv3D	该层将三维卷积应用于输入
Cropping1D	该层沿着时间维度裁剪输入数据
Cropping2D	该层沿着空间维度裁剪输入数据，例如图像中的宽度和高度
Cropping3D	该层会在时空上裁剪输入数据，即所有三个维度
UpSampling1D	该层沿着时间轴按指定的次数重复输入数据
UpSampling2D	该层沿着行和列按指定时间重复输入数据
UpSampling3D	该层沿着三个维度按指定的时间重复输入数据
ZeroPadding1D	该层将零添加到时间维度的开始和结束处
ZeroPadding2D	该层将零行和零列添加到二维张量的顶部、底部、左侧或右侧
ZeroPadding3D	该层将零张量添加到三维张量的三个维度中

3.4.3 Keras 池化层

这些层为 CNN 提供不同的池化操作，见表 3-3。

表 3-3

网络层名称	描述
MaxPooling1D	该层对一维输入数据实现最大池化操作
MaxPooling2D	该层对二维输入数据实现最大池化操作
MaxPooling3D	该层对三维输入数据实现最大池化操作
AveragePooling1D	该层对一维输入数据实现平均池化操作
AveragePooling2D	该层对二维输入数据实现平均池化操作
AveragePooling3D	该层对三维输入数据实现平均池化操作
GlobalMaxPooling1D	该层对一维输入数据实现全局最大池化操作
GlobalAveragePooling1D	该层对一维输入数据实现全局平均池化操作
GlobalMaxPooling2D	该层对二维输入数据实现全局最大池化操作
GlobalAveragePooling2D	该层对二维输入数据实现全局平均池化操作

3.4.4 Keras 局连接层

这些层在 CNN 中很有用，见表 3-4。

表 3-4

网络层名称	描述
LocallyConnected1D	该层会在单个空间或时间维度上对输入数据进行卷积，为了不共享权重，会在输入的每个块上采用不同的滤波器
LocallyConnected2D	该层会在两个维度上对输入数据进行卷积，为了不共享权重，会在输入的每个块上采用不同的滤波器

3.4.5 Keras 循环层

这些层可用来实现各种 RNN，见表 3-5。

表 3-5

网络层名称	描述
SimpleRNN	该层实现全连接的 RNN
GRU	该层实现门控循环单元网络
LSTM	该层实现了长短期记忆（LSTM）网络

3.4.6 Keras 嵌入层

目前，只有一个嵌入层可用，见表 3-6。

表 3-6

网络层名称	描述
Embedding	该层采用由索引组成的二维张量形状（batch_size, sequence_length），并产生一个密集向量组成的张量，其结构为（batch_size, sequence_length, output_dim）

3.4.7 Keras 合并层

这些层合并两个或多个输入张量，并通过应用每个层的特定操作来生成单个输出张量，见表 3-7。

表 3-7

网络层名称	描述
Add	该层对输入张量进行逐元素相加
Multiply	该层对输入张量进行逐元素相乘
Average	该层计算输入张量的平均值
Maximum	该层计算输入张量各元素最大值
Concatenate	该层将输入张量沿指定的轴连接
Dot	该层计算两个输入张量之间的内积
add、multiply、average、maximum、concatenate 和 dot	这些函数表示这里介绍的各个合并层的函数接口

3.4.8 Keras 高级激活层

这些层实现了高级激活函数，这些函数不能作为简单的底层后端函数。其操作与内核层中的 Activation( ) 层类似，见表 3-8。

表 3-8

网络层名称	描述
LeakyReLU	该层为 ReLU 激活函数的泄漏版本
PReLU	该层为参数化 ReLU 激活函数
ELU	该层为指数线性单位激活函数
ThresholdedReLU	该层为 ReLU 激活函数的阈值版本

3.4.9 Keras 归一化层

目前，只有一个归一化层，见表 3-9。

表 3-9

网络层名称	描述
BatchNormalization	该层在每个批处理中对前一层的输出进行归一化，使得该层的输出近似为具有接近零的平均值和接近于 1 的标准偏差

3.4.10 Keras 噪声层

可以将这些层添加到模型中，以便通过增加噪声来防止过度拟合，这也被称为正则化层。这些层的操作方式与内核层中的 Dropout( ) 和 ActivityRegularizer( ) 层相同，见表 3-10。

表 3-10

网络层名称	描述
GaussianNoise	该层将可加的 0 中心高斯噪声应用于输入
GaussianDropout	该层将可乘的 1 中心高斯噪声应用于输入
AlphaDropout	该层删除一定比例的输入，使得输出的均值和方差与输入的均值和方差很靠近

3.5 将网络层添加到 Keras 模型中

上一节提到的所有网络层都需要添加到之前创建的模型中。下面的章节将介绍如何使用功能性 API 和序列化 API 来添加神经网络层。

3.5.1 利用序列化 API 将网络层添加到 Keras 模型中

在序列化 API 中，可以通过实例化前面各节所给出的某种网络层的对象来创建网络层，然后使用 model.add( ) 函数将创建的网络层添加到模型中。下面将创建一个模型，然后为其添加两个网络层：

```
model = Sequential()
model.add(Dense(10, input_shape=(256,)))
model.add(Activation('tanh'))
model.add(Dense(10))
model.add(Activation('softmax'))
```

3.5.2 利用功能性 API 将网络层添加到 Keras 模型中

在功能性 API 中，首先以功能性方式创建网络层，然后在创建模型时，将输入和输出网络层作为张量的参数（这种方式在前面介绍过）。

下面是一个具体的例子。

1）首先，创建输入层：

```
input = Input(shape=(64,))
```

2）接下来，从输入层中以功能性方式创建密集层：

```
hidden = Dense(10)(inputs)
```

3）以功能性的方式从前面的网络层创建更深层的隐藏层：

```
hidden = Activation('tanh')(hidden)
hidden = Dense(10)(hidden)
output = Activation('tanh')(hidden)
```

4）最后，用输入层和输出层实例化模型对象：

```
model = Model(inputs=input, outputs=output)
```

有关创建序列化和功能性 Keras 模型的更多细节，可以阅读由 Antonio Gulli 和 Sujit Pal 编写的《Keras 深度学习》一书，该书由 Packt 于 2017 年出版。

3.6 编译 Keras 模型

前面所建立的模型，在用于训练和预测之前，需要使用 model.compile( ) 方法进行编译。compile( ) 方法的定义如下：

```
compile(self, optimizer, loss, metrics=None, sample_weight_mode=None)
```

编译方法有三个参数：

- 优化器（optimizer）：可以指定自己的函数或 Keras 提供的某个函数，用于在优化迭代中更新参数。Keras 提供以下内置的优化器函数：
  - SGD
  - RMSprop
  - Adagrad
  - Adadelta
  - Adam
  - Adamax

- Nadam
- 损失函数：可以指定自己的损失函数或使用自定义的损失函数。使用优化器函数来优化参数，从而使损失函数最小化。Keras 提供以下损失函数：
  - mean_squared_error
  - mean_absolute_error
  - mean_absolute_pecentage_error
  - mean_squared_logarithmic_error
  - squared_hinge
  - hinge
  - categorical_hinge
  - sparse_categorical_crossentropy
  - binary_crossentropy
  - poisson
  - cosine proximity
  - binary_accuracy
  - categorical_accuracy
  - sparse_categorical_accuracy
  - top_k_categorical_accuracy
  - sparse_top_k_categorical_accuracy
- 度量：第三个参数是训练模型时需要收集的度量标准列表。如果详细输出处于打开状态，则会为每次迭代打印度量信息。这些度量就像损失函数一样；Keras 可让用户自定义度量函数。所有的损失函数也可以作为度量函数使用。

3.7 训练 Keras 模型

训练 Keras 模型只需要简单调用 model.fit( ) 方法，该方法的完整定义如下：

```
fit(self, x, y, batch_size=32, epochs=10, verbose=1, callbacks=None,
    validation_split=0.0, validation_data=None, shuffle=True,
    class_weight=None, sample_weight=None, initial_epoch=0)
```

> 这里不会详细介绍该方法，读者可以从 Keras 网站 https://keras.io/models/sequential/ 上查阅该方法的详细信息。

对于之前创建的模型，可使用以下代码来训练：

```
model.fit(x_data, y_labels)
```

3.8 使用 Keras 模型进行预测

训练好的模型可用于使用 model.predict( ) 方法进行预测，或使用 model.evaluate( ) 方法评估模型。

两种方法的定义如下：

```
predict(self, x, batch_size=32, verbose=0)
evaluate(self, x, y, batch_size=32, verbose=1, sample_weight=None)
```

3.9 Keras 中的其他模块

Keras 提供了几个额外的模块，这些模块是对基本的工作流程（如本章开始部分的描述）的补充。一些模块具有如下功能：

- preprocessing 模块提供多种函数来对序列、图像和文本数据进行预处理。
- datasets 模块提供了一些函数来快速访问几种常用数据集，如 CIFAR10 图像数据集、CIFAR100 图像数据集、IMDB 电影评论数据集、路透 newswire 主题数据集、MNIST 手写数字数据集和 Boston 房价数据集。
- initializers 模块提供了一些函数来初始化网络层的随机权重参数，例如：Zeros、Ones、Constant、RandomNormal、RandomUniform、TruncatedNormal、VarianceScaling、Orthogonal、Identity、lecun_normal、lecun_uniform、glorot_normal、glorot_uniform、he_normal 和 he_uniform。
- models 模块提供了一些函数来恢复模型体系的架构和权重，例如 model_from_json、model_from_yaml 和 load_model。模型结构可以用 model.to_yaml( ) 和 model.to_json( ) 方法保存。模型权重可以用 model.save( ) 方法保存。权重被保存在一个 HDF5 文件中。
- applications 模块提供了一些预先训练好的模型，例如 Xception、VGG16、VGG19、ResNet50、Inception V3、InceptionResNet V2 和 MobileNet。下面将学习如何使用预先训练好的模型来进行预测，还将学习如何使用不同领域的数据集来重新训练 applications 模块中的那些预先训练好的模型。

现在，本章结束了对 TensorFlow 的高级框架 Keras 的简要介绍，下面将提供 Keras 构建模型的示例。

3.10 基于 MNIST 数据集的 Keras 顺序模型示例

下面将创建一个简单的多层感知器（在第 5 章中详细介绍），它用于对 MNIST 数据集中的手写数字进行分类。

```
import keras
from keras.datasets import mnist
from keras.models import Sequential
from keras.layers import Dense, Dropout
from keras.optimizers import SGD
from keras import utils
import numpy as np

# 定义一些超级参数
batch_size = 100
n_inputs = 784
n_classes = 10
```

```python
n_epochs = 10

# 得到数据
(x_train, y_train), (x_test, y_test) = mnist.load_data()

# 重塑二维 28 × 28 像素
# 将图像大小调整为784像素的单个向量
x_train = x_train.reshape(60000, n_inputs)
x_test = x_test.reshape(10000, n_inputs)

# 将输入值转换为 float32
x_train = x_train.astype(np.float32)
x_test = x_test.astype(np.float32)

# 将图像向量的值归一化为小于 1
x_train /= 255
x_test /= 255

# 将输出数据转换为独热编码格式
y_train = utils.to_categorical(y_train, n_classes)
y_test = utils.to_categorical(y_test, n_classes)

# 构建序列化模型
model = Sequential()
# 第一层必须指定输入向量的大小
model.add(Dense(units=128, activation='sigmoid', input_shape=(n_inputs,)))
# 添加dropout层以防止过度拟合
model.add(Dropout(0.1))
model.add(Dense(units=128, activation='sigmoid'))
model.add(Dropout(0.1))
# 输出层只能使神经元等于输出数量
model.add(Dense(units=n_classes, activation='softmax'))

# 输出模型的摘要
model.summary()

# 编译模型
model.compile(loss='categorical_crossentropy',
              optimizer=SGD(),
              metrics=['accuracy'])

# 训练模型
model.fit(x_train, y_train,
          batch_size=batch_size,
          epochs=n_epochs)

# 评估模型并输出准确度分数
scores = model.evaluate(x_test, y_test)

print('\n loss:', scores[0])
print('\n accuracy:', scores[1])
```

从描述和训练的 Keras 模型可以获得以下输出结果:

```
Layer (type)                 Output Shape              Param #
=================================================================
dense_7 (Dense)              (None, 128)               100480
_____
dropout_5 (Dropout)          (None, 128)               0
_____
dense_8 (Dense)              (None, 128)               16512
_____
dropout_6 (Dropout)          (None, 128)               0
_____
dense_9 (Dense)              (None, 10)                1290
=================================================================
Total params: 118,282
Trainable params: 118,282
Non-trainable params: 0

Epoch 1/10
60000/60000 [==============================] - 3s - loss: 2.3018 - acc: 0.1312
Epoch 2/10
60000/60000 [==============================] - 2s - loss: 2.2395 - acc: 0.1920
Epoch 3/10
60000/60000 [==============================] - 2s - loss: 2.1539 - acc: 0.2843
Epoch 4/10
60000/60000 [==============================] - 2s - loss: 2.0214 - acc: 0.3856
Epoch 5/10
60000/60000 [==============================] - 3s - loss: 1.8269 - acc: 0.4739
Epoch 6/10
60000/60000 [==============================] - 2s - loss: 1.5973 - acc: 0.5426
Epoch 7/10
60000/60000 [==============================] - 2s - loss: 1.3846 - acc: 0.6028
Epoch 8/10
60000/60000 [==============================] - 3s - loss: 1.2133 - acc: 0.6502
Epoch 9/10
60000/60000 [==============================] - 3s - loss: 1.0821 - acc: 0.6842
Epoch 10/10
60000/60000 [==============================] - 3s - loss: 0.9799 - acc: 0.7157

loss: 0.859834249687
accuracy: 0.788
```

可以看到在 Keras 中构建和训练模型会很容易。

可以从网站 https://keras.io 中获得更多有关 Keras 的信息。

3.11 总结

本章介绍了 Keras。它是 TensorFlow 最受欢迎的高级库。笔者更喜欢将 Keras 用于开发商业模型，有时也用于学术研究。本章介绍了 Keras 的工作流程，以及如何使用功能性

API 和序列化 API 来创建和训练模型。本章也介绍了各种 Keras 网络层以及如何将这些层添加到序列模型和功能模型中，还学习如何编译、训练和评估 Keras 模型。另外，也介绍了 Keras 所提供的一些附加模块。

本书后面的章节将涵盖 TensorFlow 内核和 Keras 中的大部分示例。在下一章中，将学习如何使用 TensorFlow 来构建分类和回归模型，这些属于传统机器学习模型。

第 4 章 基于 TensorFlow 的经典机器学习算法

机器学习是计算机科学的一个领域，涉及算法的研究、开发和应用，它能够让计算机通过数据来学习。计算机学习的模型被用来进行预测。机器学习研究人员和工程师通过构建模型，然后使用这些模型进行预测来达到此目标。现在，机器学习在自然语言理解、视频处理、图像识别、语音和计算机视觉等很多领域得到了广泛的应用。

下面介绍什么是模型。所有机器学习的问题都可以被抽象为如下等式：

$$y=f(x)$$

式中，y 是输出（或目标）；x 是输入（或特征）。如果 x 是特征集合，也称其为特征向量，并用 X 表示。所谓模型，就是指找到将特征映射到目标的函数 f。因此，一旦找到 f，就可预测新值 x 的输出值 y。

机器学习的核心是找到函数 f，该函数可以根据 x 的值来预测 y。例如，回想一下初中数学中的线性方程：

$$y=mx+c$$

可将上面的简单等式重写为：

$$y=Wx+b$$

式中，W 被称为权重；b 被称为偏差。不要担心权重和偏差这些术语，稍后将会介绍。到目前为止，可以把 W 看作线性方程中的 m，b 看作线性方程中的 c。因此，机器学习问题可以被描述为从当前的 X 中找出 W 和 b，使其可以预测 y 的值。

回归分析或回归建模是指用于估计变量之间关系的方法和技术。回归模型的输入变量称为独立变量（或预测变量、特征），而回归模型的输出变量称为因变量或目标。回归模型定义如下：

$$y \approx f(X, \beta)$$

式中，y 是目标变量；X 是特征向量；β 是参数向量。

 通常，人们使用一种非常简单的回归形式（即线性回归）来估计参数 β。

在机器学习中，必须要从给定的数据中学习模型的参数 β_0 和 β_1，这是为了能通过将来的 X 值来预测 Y 值。称权重为 β_1，偏差为 β_0，并分别用 w 和 b 来表示它们。

因此模型变成如下：

$$y \approx X \times w + b$$

分类是机器学习中的经典问题之一。分类需要考虑数据可能属于某个类别，例如，如果提供的是图像数据，则可能是猫或狗的图片。因此，在这种情况下，类别就是猫和狗。分类意味着对正在观察的数据或对象的标签进行识别。分类属于监督机器学习的范畴。在分类问题中，提供的数据集有特征和与特征相对应的类别标签。通过训练数据集得到的模型称为训练模型，这其实是通过计算来得到模型的参数。训练好的模型可以用于识别新数据的类标签。

分类问题可以有两种类型：二分类和多分类。二分类是指将数据分为两个不同的离散标签，例如，患有癌症的病人或没有患癌症的病人，图像是猫还是狗。多分类意味着数据将被分类到多个类别中，例如电子邮件分类问题会将电子邮件分成社交媒体电子邮件、与工作有关的电子邮件、个人电子邮件、与家人有关的电子邮件、垃圾邮件、购物电子邮件等。另一个例子是数字图片的类别问题，每张图片可能具有 0 ~ 9 之间的标记，具体取决于为图片标记的数字。本章将给出二分类以及多分类问题的示例。

本章将进一步讨论以下主题：
- 回归：
  - 简单的线性回归；
  - 多元回归；
  - 正则化回归；
  - Lasso 正则化；
  - 岭正则化；
  - 弹性网正则化。
- 分类：
  - 使用 Logistic 回归进行分类；
  - 二分类；
  - 多分类。

4.1 简单的线性回归

读者可能使用过其他机器学习库。下面将介绍如何使用 TensorFlow 来学习简单的线性回归模型。在谈到特定领域的示例之前，先使用生成的数据集来解释相关概念。

这里将使用生成的数据集，以便不同领域的读者都能够学习，而不会对特定领域示例中的细节感到困惑。

可以查看 Jupyter 笔记本 ch-04a_Regression 中的代码。

4.1.1 数据准备

用 make_regression 函数（它在 sklearn 库的 dataset 模块中）来生成数据集：

```
from sklearn import datasets as skds
X, y = skds.make_regression(n_samples=200,
                            n_features=1,
                            n_informative=1,
                            n_targets=1,
                            noise = 20.0)
```

上述函数将生成一个用于建立回归模型的数据集，它包含 200 个样本，每个样本有一个特征和一个目标值，并且样本包含一些噪声。由于只生成一个目标值，可使用一维 NumPy 数组来生成 y；可重塑 y 来得到两维：

```
if (y.ndim == 1):
    y = y.reshape(len(y),1)
```

使用以下代码来绘制并查看生成的数据集：

```
import matplotlib.pyplot as plt
plt.figure(figsize=(14,8))
plt.plot(X,y,'b.')
plt.title('Original Dataset')
plt.show()
```

运行上面的代码将得到图 4-1 的结果，由于生成的数据是随机的，每次可能会得到不同的结果。

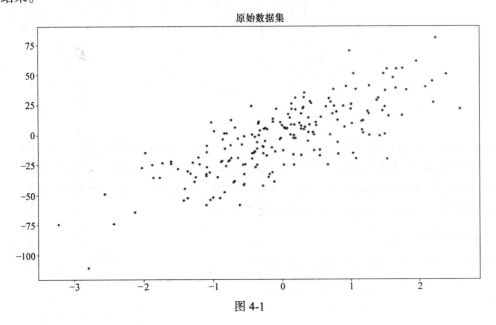

图 4-1

现在把数据分成训练集和测试集：

```
X_train, X_test, y_train, y_test = skms.train_test_split(X, y,
                                            test_size=.4,
                                            random_state=123)
```

4.1.2 建立简单的回归模型

通常，在 TensorFlow 中构建和训练回归模型包括以下步骤：
1）定义输入、参数和其他变量；
2）定义模型；
3）定义损失函数；
4）定义优化器；
5）训练模型进行一系列称为 epochs 的迭代。

4.1.2.1 定义输入、参数和其他变量

在使用 TensorFlow 构建和训练回归模型之前，需要定义一些重要的变量和操作。从 x_train 和 y_train 中找出输出和输入变量的数量，然后使用这些数字来定义 x（x_tensor）、y（y_tensor）、权重（w）和偏差（b）。

```
num_outputs = y_train.shape[1]
num_inputs = X_train.shape[1]

x_tensor = tf.placeholder(dtype=tf.float32,
                   shape=[None, num_inputs],
                   name="x")
y_tensor = tf.placeholder(dtype=tf.float32,
                   shape=[None, num_outputs],
                   name="y")

w = tf.Variable(tf.zeros([num_inputs,num_outputs]),
             dtype=tf.float32,
             name="w")
b = tf.Variable(tf.zeros([num_outputs]),
             dtype=tf.float32,
             name="b")
```

- x_tensor 的行数没有确定，列数为 num_inputs，在这个示例中列的数量只有 1。
- y_tensor 的行数没有确定，列数为 num_outputs，在这个示例中列的数量只有 1。
- w 的维数为 num_inputs 乘以 num_outputs，在这个示例中，w 的维数为 1×1。
- b 的维数为 num_outputs，在这个示例中 b 为 1。

4.1.2.2 定义模型

接下来，将模型定义为 (x_tensor × w) + b：

```
model = tf.matmul(x_tensor, w) + b
```

4.1.2.3 定义损失函数

接下来,使用均方误差(mse)作为损失函数。均方误差定义如下:

$$\frac{1}{n}\sum(y_i - \hat{y}_i)^2$$

有关均方误差的详细信息可从以下链接找到:
https://en.wikipedia.org/wiki/Mean_squared_error;
http://www.statisticshowto.com/mean-squared-error/。

y 的实际值与估计值的差别被称为残差(residual)。损失函数计算的是残差的平方的均值,在 TensorFlow 中用以下方式定义:

```
loss = tf.reduce_mean(tf.square(model - y_tensor))
```

- model - y_tensor 计算残差。
- tf.square(model - y_tensor) 计算每个残差的平方。
- tf.reduce_mean(...) 计算上一步结果的平均值,作为最终结果。

也可以定义均方误差(mean squared error,mse)和 r 平方(r-squared,rs)函数来评估训练后的模型。因为在接下来的章节中,损失函数将会改变,但均方误差函数将保持不变,所以会使用单独的均方误差函数。

```
# mse 和 R2 函数
mse = tf.reduce_mean(tf.square(model - y_tensor))
y_mean = tf.reduce_mean(y_tensor)
total_error = tf.reduce_sum(tf.square(y_tensor - y_mean))
unexplained_error = tf.reduce_sum(tf.square(y_tensor - model))
rs = 1 - tf.div(unexplained_error, total_error)
```

4.1.2.4 定义优化器函数

接下来,以 0.001 的学习率来实例化 theGradientDescentOptimizer 函数,并通过它来最小化损失函数:

```
learning_rate = 0.001
optimizer = tf.train.GradientDescentOptimizer(learning_rate).minimize(loss)
```

有关梯度下降的详细信息可从以下链接找到:
https://en.wikipedia.org/wiki/Gradient_descent;
https://www.analyticsvidhya.com/blog/2017/03/introduction-to-gradient-descent-algorithm-along-its-variants/。

TensorFlow 提供了许多其他的优化函数,比如 Adadelta、Adagrad 和 Adam。后面的章节将对其中的一些函数进行介绍。

4.1.2.5 训练模型

现在定义了模型、损失函数和优化器,以便对模型进行训练来得到参数 w 和 b。为了训练模型,定义以下全局变量:

- num_epochs:运行训练程序的迭代次数,每次迭代,模型都会学习更好的参数,这点将在后面的图中看到。
- w_hat 和 b_hat:通过估计得到的参数 w 和 b。
- loss_epochs、mse_epochs、rs_epochs:要在每次迭代中收集训练数据集上的总误差值,以及基于测试数据集上模型的均方误差和 r 平方值。
- mse_score 和 rs_score:收集最终训练模型的均方误差和 r 平方值。

```
num_epochs = 1500
w_hat = 0
b_hat = 0
loss_epochs = np.empty(shape=[num_epochs],dtype=float)
mse_epochs = np.empty(shape=[num_epochs],dtype=float)
rs_epochs = np.empty(shape=[num_epochs],dtype=float)

mse_score = 0
rs_score = 0
```

在完成对会话和全局变量的初始化后,需要运行 num_epoch 次循环:

```
with tf.Session() as tfs:
    tf.global_variables_initializer().run()
    for epoch in range(num_epochs):
```

每次迭代都会在训练数据上进行优化:

```
tfs.run(optimizer, feed_dict={x_tensor: X_train, y_tensor: y_train})
```

为了绘图,需使用学到的 w 和 b 值来计算误差,并将结果保存在 loss_val 中:

```
loss_val = tfs.run(loss,feed_dict={x_tensor: X_train, y_tensor: y_train})
loss_epochs[epoch] = loss_val
```

计算测试数据的预测值的均方误差和 r 平方值:

```
mse_score = tfs.run(mse,feed_dict={x_tensor: X_test, y_tensor: y_test})
mse_epochs[epoch] = mse_score

rs_score = tfs.run(rs,feed_dict={x_tensor: X_test, y_tensor: y_test})
rs_epochs[epoch] = rs_score
```

最后,一旦循环结束,保存 w 和 b 值,用于之后的绘图:

```
w_hat,b_hat = tfs.run([w,b])
w_hat = w_hat.reshape(1)
```

在 2000 次迭代后输出模型和测试数据上的最终均方误差:

```
print('model : Y = {0:.8f} X + {1:.8f}'.format(w_hat[0],b_hat[0]))
print('For test data : MSE = {0:.8f}, R2 = {1:.8f} '.format(
    mse_score,rs_score))
```

得到以下输出结果：

```
model : Y = 20.37448120 X + -2.75295663
For test data : MSE = 297.57995605, R2 = 0.66098368
```

从该结果可看出：训练得到的模型并不是一个很好的模型，但在后面的章节中将会看到如何使用神经网络来改进它。

本章的目标是介绍如何在不使用神经网络的情况下通过 TensorFlow 来构建和训练回归模型。

下面来绘制估计的模型和原始数据：

```
plt.figure(figsize=(14,8))
plt.title('Original Data and Trained Model')
x_plot = [np.min(X)-1,np.max(X)+1]
y_plot = w_hat*x_plot+b_hat
plt.axis([x_plot[0],x_plot[1],y_plot[0],y_plot[1]])
plt.plot(X,y,'b.',label='Original Data')
plt.plot(x_plot,y_plot,'r-',label='Trained Model')
plt.legend()
plt.show()
```

得到如图 4-2 所示的原始数据和训练模型的示意图。

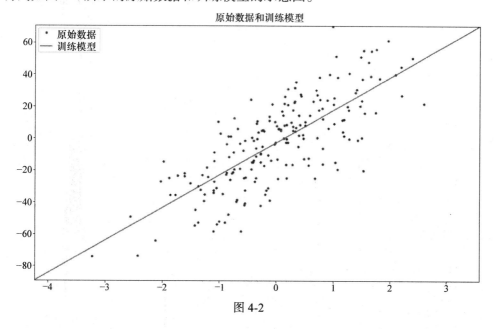

图 4-2

在每次迭代中绘制训练数据和测试数据的均方误差：

```
plt.figure(figsize=(14,8))

plt.axis([0,num_epochs,0,np.max(loss_epochs)])
plt.plot(loss_epochs, label='Loss on X_train')
plt.title('Loss in Iterations')
plt.xlabel('# Epoch')
plt.ylabel('MSE')
```

```
plt.axis([0,num_epochs,0,np.max(mse_epochs)])
plt.plot(mse_epochs, label='MSE on X_test')
plt.xlabel('# Epoch')
plt.ylabel('MSE')
plt.legend()

plt.show()
```

得到的结果如图 4-3 所示，该图显示每次迭代时，均方误差会减小，然后保持在 500 附近的水平线上。

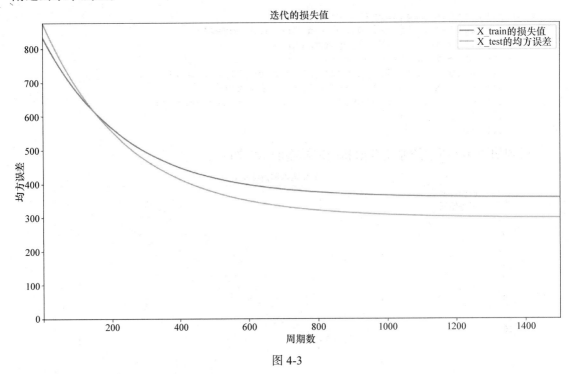

图 4-3

绘制 r 平方值：

```
plt.figure(figsize=(14,8))
plt.axis([0,num_epochs,0,np.max(rs_epochs)])
plt.plot(rs_epochs, label='R2 on X_test')
plt.xlabel('# Epoch')
plt.ylabel('R2')
plt.legend()
plt.show()
```

当绘制迭代步数与 r 平方值时，会得到如图 4-4 所示的结果。

图 4-4 表明模型从非常低的 r 平方值开始，但是随着不断训练模型，其误差会减小，r 平方值开始变得更大并且最终稳定在略大于 0.6 的地方。

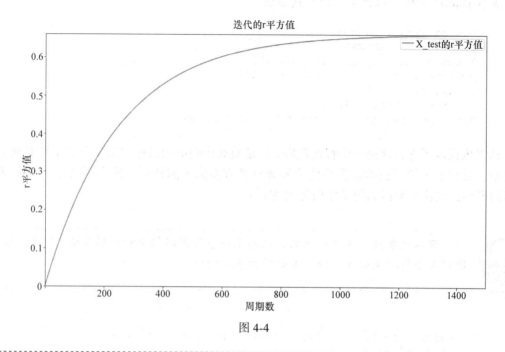

图 4-4

绘制均方误差和 r 平方值可以看到模型如何能快速地被训练，以及何时开始变得稳定，从中可以看出：进一步的训练对减少误差几乎没有任何好处。

4.1.3 使用训练好的模型进行预测

在训练好模型之后，可以用它来预测新数据。由于直线可能无法完美地拟合数据，线性模型的预测是在理解上图中一些最小均方误差的情况下进行的。

为了得到更好的拟合模型，必须使用其他方法来扩展模型，例如添加变量的线性组合。

4.2 多元回归

现在已经学会了如何使用 TensorFlow 创建一个基本的回归模型。下面尝试在不同的示例数据集上运行该模型。前面生成的数据集是单变量，即目标仅依赖于一个特征。

实际上，大多数数据集都是多变量的，而且目标是依赖于多个变量或特征的，因此回归模型被称为多元回归（或多维回归）。

首先从最流行的 Boston 数据集开始。该数据集包含 Boston 506 所房屋的 13 个属性，例如平均每间住房的房间数量、一氧化氮浓度、与 5 个 Boston 就业中心的加权距离等。输出的目标是房屋价值的中位值。下面来深入研究这个数据集的回归模型。

加载 sklearn 库中的数据集并查看其描述：

```
boston=skds.load_boston()
print(boston.DESCR)
X=boston.data.astype(np.float32)
y=boston.target.astype(np.float32)
if (y.ndim == 1):
    y = y.reshape(len(y),1)
X = skpp.StandardScaler().fit_transform(X)
```

这里也提取了 X，这是一个特征矩阵，y 是前面代码中的目标向量。重塑 y 使其成为二维向量，并将 x 中的特征缩放为均值为零和标准方差为 1 的特征。现在用这个 X 和 y 来训练回归模型，就像在前面的例子中所做的那样：

可以观察到，这个例子的代码与上一节简单回归中的代码类似；但是，这里使用多个特征来训练模型，因此称为多元回归。

```
X_train, X_test, y_train, y_test = skms.train_test_split(X, y,
    test_size=.4, random_state=123)
num_outputs = y_train.shape[1]
num_inputs = X_train.shape[1]

x_tensor = tf.placeholder(dtype=tf.float32,
    shape=[None, num_inputs], name="x")
y_tensor = tf.placeholder(dtype=tf.float32,
    shape=[None, num_outputs], name="y")

w = tf.Variable(tf.zeros([num_inputs,num_outputs]),
    dtype=tf.float32, name="w")
b = tf.Variable(tf.zeros([num_outputs]),
    dtype=tf.float32, name="b")

model = tf.matmul(x_tensor, w) + b
loss = tf.reduce_mean(tf.square(model - y_tensor))
# mse 和 R2 函数
mse = tf.reduce_mean(tf.square(model - y_tensor))
y_mean = tf.reduce_mean(y_tensor)
total_error = tf.reduce_sum(tf.square(y_tensor - y_mean))
unexplained_error = tf.reduce_sum(tf.square(y_tensor - model))
rs = 1 - tf.div(unexplained_error, total_error)

learning_rate = 0.001
optimizer = tf.train.GradientDescentOptimizer(learning_rate).minimize(loss)

num_epochs = 1500
loss_epochs = np.empty(shape=[num_epochs],dtype=np.float32)
mse_epochs = np.empty(shape=[num_epochs],dtype=np.float32)
```

```
        rs_epochs = np.empty(shape=[num_epochs],dtype=np.float32)

        mse_score = 0
        rs_score = 0

        with tf.Session() as tfs:
            tfs.run(tf.global_variables_initializer())
            for epoch in range(num_epochs):
                feed_dict = {x_tensor: X_train, y_tensor: y_train}
                loss_val, _ = tfs.run([loss, optimizer], feed_dict)
                loss_epochs[epoch] = loss_val

                feed_dict = {x_tensor: X_test, y_tensor: y_test}
                mse_score, rs_score = tfs.run([mse, rs], feed_dict)
                mse_epochs[epoch] = mse_score
                rs_epochs[epoch] = rs_score

        print('For test data : MSE = {0:.8f}, R2 = {1:.8f} '.format(
            mse_score, rs_score))
```

从模型中获得以下输出：

```
For test data : MSE = 30.48501778, R2 = 0.64172244
```

下面来绘制均方误差和 r 平方值。

图 4-5 为均方误差的示意图。

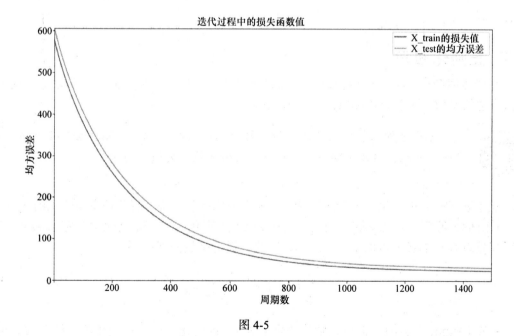

图 4-5

图 4-6 为 r 平方值的示意图。

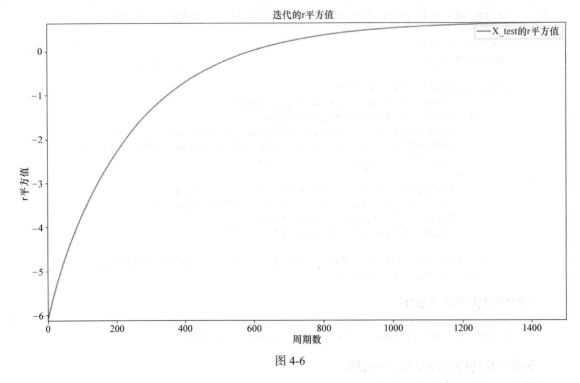

图 4-6

正如绘制单变量数据集时所看到的,可以看到在多变量情况下,均方误差和 r 平方值也有类似情况。

4.3 正则化回归

在线性回归中,训练模型会返回训练数据上的最佳拟合参数。但是,在训练数据中找到最合适的参数可能会导致过拟合(overfitting)。

 过拟合意味着该模型能很好地拟合训练数据,但会给测试数据带来更大的误差。因此,通常为模型添加一个惩罚项来获得更简单的模型。

这个惩罚称为正则化,这种回归模型称为正则化回归模型。有三种主要的正则化模型:

- Lasso 回归:在 Lasso 正则化(也称为 L1 正则化)中,正则化项是指 Lasso 参数 α 乘以权重向量 w 的各元素的绝对值之和。因此,损失函数如下:

$$\frac{1}{n}\sum_{i=1}^{n}(y_i - \hat{y}_i)^2 + \alpha \frac{1}{n}\sum_{i=1}^{n}|w_i|$$

- 岭(Ridge)回归:在岭正则化(也称为 L2 正则化)中,正则化项是岭参数 α 乘以权重向量 w 的各元素的平方之和。因此,损失函数如下:

$$\frac{1}{n}\sum_{i=1}^{n}(y_i-\hat{y}_i)^2+\alpha\frac{1}{n}\sum_{i=1}^{n}w_i^2$$

- **弹性网（ElasticNet）回归**：把 Lasso 正则化项和岭正则化项组合在一起，所得到的正则化被称为弹性网正则化。因此，损失函数为

$$\frac{1}{n}\sum_{i=1}^{n}(y_i-\hat{y}_i)^2+\alpha_1\frac{1}{n}\sum_{i=1}^{n}|w_i|+\alpha_2\frac{1}{n}\sum_{i=1}^{n}w_i^2$$

有关正规化的更多信息，请参阅以下资源：http://www.statisticshowto.com/regularization/。

当想要删除某些特征时，一个简单的经验法则是使用 L1 或 Lasso 正则项，这样会减少计算时间，但同时也会降低准确性。

现在来看看在 TensorFlow 中如何实现这些正则化损失函数。这里将继续使用前面例子中所使用的 Boston 数据集。

4.3.1 Lasso 正则化

定义 Lasso 参数的值为 0.8：

```
lasso_param = tf.Variable(0.8, dtype=tf.float32)
lasso_loss = tf.reduce_mean(tf.abs(w)) * lasso_param
```

将 Lasso 参数设置为零表示没有正则化，正规化项的值越高，惩罚越高。以下是 Lasso 正则化回归的完整代码，这些代码会通过训练模型来预测 Boston 的房价：

下面的代码假设训练数据集和测试数据集会按照前面的例子进行拆分。

```
num_outputs = y_train.shape[1]
num_inputs = X_train.shape[1]

x_tensor = tf.placeholder(dtype=tf.float32,
                shape=[None, num_inputs], name='x')
y_tensor = tf.placeholder(dtype=tf.float32,
                shape=[None, num_outputs], name='y')

w = tf.Variable(tf.zeros([num_inputs, num_outputs]),
            dtype=tf.float32, name='w')
b = tf.Variable(tf.zeros([num_outputs]),
            dtype=tf.float32, name='b')

model = tf.matmul(x_tensor, w) + b

lasso_param = tf.Variable(0.8, dtype=tf.float32)
lasso_loss = tf.reduce_mean(tf.abs(w)) * lasso_param
```

```python
loss = tf.reduce_mean(tf.square(model - y_tensor)) + lasso_loss

learning_rate = 0.001
optimizer = tf.train.GradientDescentOptimizer(learning_rate).minimize(loss)

mse = tf.reduce_mean(tf.square(model - y_tensor))
y_mean = tf.reduce_mean(y_tensor)
total_error = tf.reduce_sum(tf.square(y_tensor - y_mean))
unexplained_error = tf.reduce_sum(tf.square(y_tensor - model))
rs = 1 - tf.div(unexplained_error, total_error)
num_epochs = 1500
loss_epochs = np.empty(shape=[num_epochs],dtype=np.float32)
mse_epochs = np.empty(shape=[num_epochs],dtype=np.float32)
rs_epochs = np.empty(shape=[num_epochs],dtype=np.float32)

mse_score = 0.0
rs_score = 0.0

num_epochs = 1500
loss_epochs = np.empty(shape=[num_epochs], dtype=np.float32)
mse_epochs = np.empty(shape=[num_epochs], dtype=np.float32)
rs_epochs = np.empty(shape=[num_epochs], dtype=np.float32)

mse_score = 0.0
rs_score = 0.0

with tf.Session() as tfs:
    tfs.run(tf.global_variables_initializer())
    for epoch in range(num_epochs):
        feed_dict = {x_tensor: X_train, y_tensor: y_train}
        loss_val,_ = tfs.run([loss,optimizer], feed_dict)
        loss_epochs[epoch] = loss_val

        feed_dict = {x_tensor: X_test, y_tensor: y_test}
        mse_score,rs_score = tfs.run([mse,rs], feed_dict)
        mse_epochs[epoch] = mse_score
        rs_epochs[epoch] = rs_score

print('For test data : MSE = {0:.8f}, R2 = {1:.8f} '.format(
    mse_score, rs_score))
```

得到如下的输出结果:

```
For test data : MSE = 30.48978233, R2 = 0.64166653
```

使用以下代码来绘制均方误差和 r 平方值:

```python
plt.figure(figsize=(14,8))

plt.axis([0,num_epochs,0,np.max([loss_epochs,mse_epochs])])
plt.plot(loss_epochs, label='Loss on X_train')
plt.plot(mse_epochs, label='MSE on X_test')
plt.title('Loss in Iterations')
plt.xlabel('# Epoch')
```

```
plt.ylabel('Loss or MSE')
plt.legend()

plt.show()

plt.figure(figsize=(14,8))

plt.axis([0,num_epochs,np.min(rs_epochs),np.max(rs_epochs)])
plt.title('R-squared in Iterations')
plt.plot(rs_epochs, label='R2 on X_test')
plt.xlabel('# Epoch')
plt.ylabel('R2')
plt.legend()

plt.show()
```

图 4-7 为损失函数的结果。

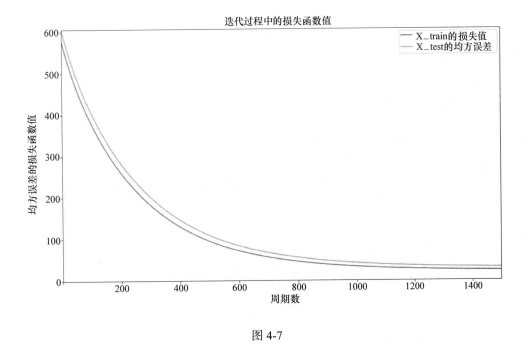

图 4-7

图 4-8 给出 r 平方值随迭代步数的变化情况。

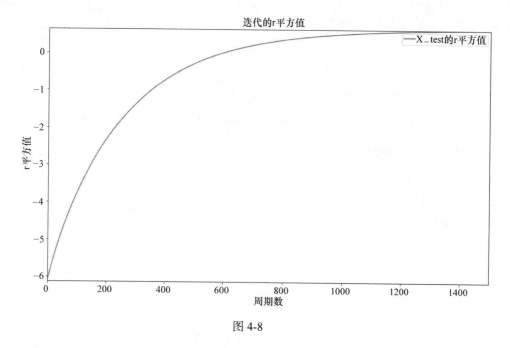

图 4-8

现在，使用岭正则化方法来重复上述示例。

4.3.2 岭正则化

以下代码是岭正则回归的完整代码，用于训练能预测 Boston 房价的模型。

```
num_outputs = y_train.shape[1]
num_inputs = X_train.shape[1]

x_tensor = tf.placeholder(dtype=tf.float32,
                          shape=[None, num_inputs], name='x')
y_tensor = tf.placeholder(dtype=tf.float32,
                          shape=[None, num_outputs], name='y')

w = tf.Variable(tf.zeros([num_inputs, num_outputs]),
                dtype=tf.float32, name='w')
b = tf.Variable(tf.zeros([num_outputs]),
                dtype=tf.float32, name='b')

model = tf.matmul(x_tensor, w) + b

ridge_param = tf.Variable(0.8, dtype=tf.float32)
ridge_loss = tf.reduce_mean(tf.square(w)) * ridge_param

loss = tf.reduce_mean(tf.square(model - y_tensor)) + ridge_loss

learning_rate = 0.001
optimizer = tf.train.GradientDescentOptimizer(learning_rate).minimize(loss)

mse = tf.reduce_mean(tf.square(model - y_tensor))
```

```
y_mean = tf.reduce_mean(y_tensor)
total_error = tf.reduce_sum(tf.square(y_tensor - y_mean))
unexplained_error = tf.reduce_sum(tf.square(y_tensor - model))
rs = 1 - tf.div(unexplained_error, total_error)

num_epochs = 1500
loss_epochs = np.empty(shape=[num_epochs],dtype=np.float32)
mse_epochs = np.empty(shape=[num_epochs],dtype=np.float32)
rs_epochs = np.empty(shape=[num_epochs],dtype=np.float32)

mse_score = 0.0
rs_score = 0.0

with tf.Session() as tfs:
    tfs.run(tf.global_variables_initializer())
    for epoch in range(num_epochs):
        feed_dict = {x_tensor: X_train, y_tensor: y_train}
        loss_val, _ = tfs.run([loss, optimizer], feed_dict=feed_dict)
        loss_epochs[epoch] = loss_val

        feed_dict = {x_tensor: X_test, y_tensor: y_test}
        mse_score, rs_score = tfs.run([mse, rs], feed_dict=feed_dict)
        mse_epochs[epoch] = mse_score
        rs_epochs[epoch] = rs_score

print('For test data : MSE = {0:.8f}, R2 = {1:.8f} '.format(
    mse_score, rs_score))
```

得到如下结果：

```
For test data : MSE = 30.64177132, R2 = 0.63988018
```

下面来绘制损失函数和均方误差的值，其中，绘制损失函数值的结果如图 4-9 所示。

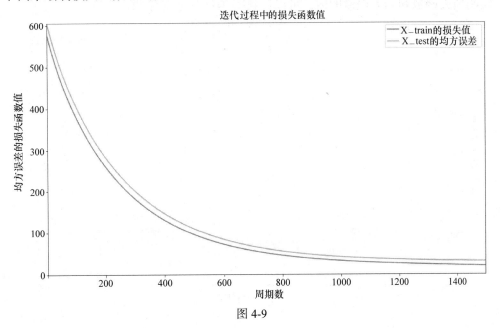

图 4-9

r平方值的情况如图4-10所示。

下面介绍Lasso正则化和岭正则化相结合的结果。

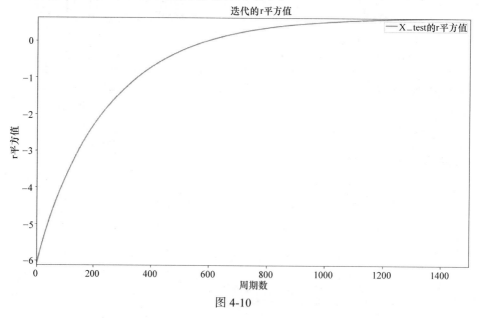

图4-10

4.3.3 弹性网正则化

ch-04a_Regression提供了弹性网正则化回归的完整代码，该代码通过训练模型来预测Boston的房价。运行该模型，可以得到以下结果：

```
For test data : MSE = 30.64861488, R2 = 0.63979971
```

绘制损失函数和均方误差的值，可以得到图4-11所示的结果。

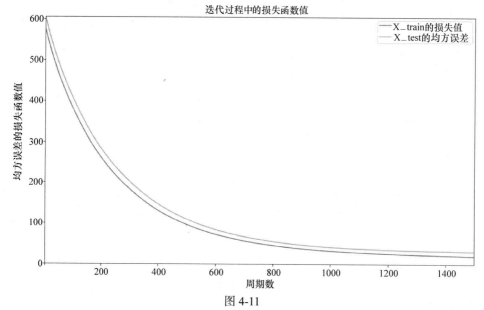

图4-11

图 4-12 是 r 平方值的绘制结果。

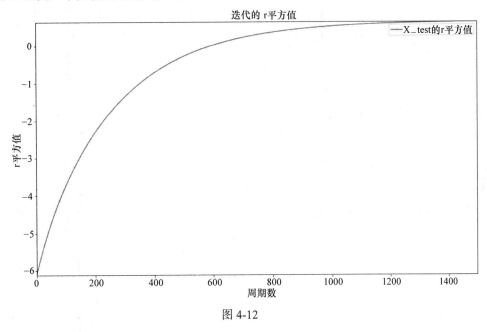

图 4-12

4.4 使用 Logistic 回归进行分类

最常见的分类方法是 Logistic 回归。Logistic 回归是一种基于概率的线性分类器。特征向量属于哪一类的概率可由下式来描述：
$$P(Y=i|x,w,b)=\phi(z)$$
其中，Y 代表输出；i 代表第 i 类；x 代表输入；w 代表权重；b 代表偏差；z 代表回归方程，有 $z=w \times x+b$；ϕ 在本示例中表示平滑函数（也称模型）。

上式表示：当给定 w 和 b 时，x 属于类别 i 的概率由函数 $\phi(z)$ 计算得到。因此，必须对模型进行训练以最大化概率值。

4.4.1 二分类的 Logistic 回归

对于二分类，函数 $\phi(z)$ 为 sigmoid 函数：
$$\phi(z)=\frac{1}{1+\mathrm{e}^{-z}}=\frac{1}{1+\mathrm{e}^{-z(w \times x+b)}}$$

sigmoid 函数产生的 y 值介于 [0，1] 之间。因此，可以使用 $y=\phi(z)$ 的值来预测类别：如果 $y>0.5$，那么类别为 1，否则类别为 0。

正如在本章的前几节中看到的那样，对于线性回归，可以采用使损失函数最小化的方式来训练模型，并且损失函数可以是平方误差或均方误差的总和。对于 Logistic 回归，希望最大化似然估计：$L(w)=P(y|x,w,b)$。

由于最大化对数似然函数更容易，故使用对数似然函数 $l(w)$ 作为损失函数。因此，损失函数（$J(w)$）被写为 $-l(w)$，可以使用像梯度下降这样的优化算法来获取该函数的最小值。

二元 Logistic 回归的损失函数可写成：

$$J(w) = -\sum_{i=1}^{n}[(y_i \times \log(\phi(z_i))) + ((1-y_i) \times (1-\log(\phi(z_i))))]$$

其中，$\phi(z)$ 是 sigmoid 函数。

下一节将介绍这种损失函数。

4.4.2 多类分类的 Logistic 回归

当涉及两个以上的类别时，Logistic 回归就被称为多分类 Logistic 回归。在多分类 Logistic 回归中，使用 softmax 函数而不是 sigmoid 函数。softmax 是最常用的函数之一：

$$\text{softmax}(\ ) = \frac{e_i^z}{\sum_j e_j^z} = \frac{e_i^{(w \times x + b)}}{\sum_j e_j^{(w \times x + b)}}$$

softmax 函数产生每个类别的概率，并且概率向量的所有元素相加为 1。当预测时，具有最高 softmax 值的类别成为输出或预测类别。正如之前讨论的那样，损失函数是负对数似然函数 $-l(w)$，可以通过优化方法（如梯度下降）来获得该函数的最小值。

多分类 Logistic 回归的损失函数可写为

$$J(w) = \sum_{i=1}^{n}[y_i \times \log(\phi(z_i))]$$

其中，$\phi(z)$ 是 softmax 函数。

本章后面将介绍该损失函数。下面来深入探讨一些例子。

可以参考 Jupyter 笔记本 ch-04b_Classification 中的一些代码。

4.5 二分类

二分类是指只有两个不同类别的问题，正如前面介绍的那样，将使用 SciKit Learn 库中的 make_classification( ) 函数来生成数据集。

```
X, y = skds.make_classification(n_samples=200,
    n_features=2,
    n_informative=2,
    n_redundant=0,
    n_repeated=0,
    n_classes=2,
    n_clusters_per_class=1)
if (y.ndim == 1):
    y = y.reshape(-1,1)
```

make_classification( ) 的参数含义非常清楚；n_samples 是要生成的数据样本的数量；n_features 是要生成的特征数量；n_classes 是类别的数量，这里为 2：

- n_samples 是要生成的数据样本的数量。为了让数据集较小，将其设为 200。
- n_features 是要生成的特征数量；为了理解 TensorFlow，需要一个简单的例子，因此这里只使用 2 个特征。
- n_classes 是类别的数量，这里取 2，因此这是一个二分类问题。

使用以下代码绘制数据：

```
plt.scatter(X[:,0],X[:,1],marker='o',c=y)
plt.show()
```

将得到图 4-13 所示结果，可能会得到一个不同的图形，因为每次运行数据生成函数时都是随机生成数据。

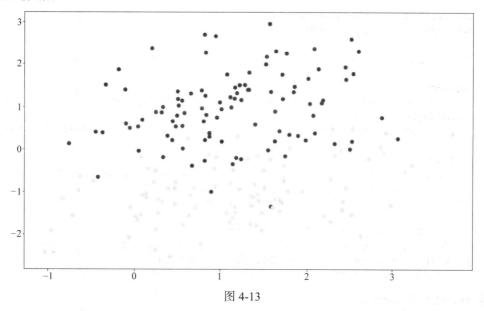

图 4-13

然后使用 NumPy 的 eye 函数将 y 转换为独热编码目标（one-hot encoded targets）类别：

```
print(y[0:5])
y=np.eye(num_outputs)[y]
print(y[0:5])
```

独热编码目标类别如下所示：

```
[1 0 0 1 0]
[[ 0.  1.]
 [ 1.  0.]
 [ 1.  0.]
 [ 0.  1.]
 [ 1.  0.]]
```

将数据分成训练数据和测试数据：

```
X_train, X_test, y_train, y_test = skms.train_test_split(
    X, y, test_size=.4, random_state=42)
```

在分类时，使用 sigmoid 函数来量化模型的值，使得输出值位于范围 [0，1] 之间。下式中的 $\phi(z)$ 为 sigmod 函数，其中 $z=w \times x+b$。损失函数现在变为由 $J(\theta)$ 表示的值，其中 θ 为参数。

$$z_i = w_i \times x_i + b$$

$$\phi(z) = \frac{1}{1+e^{-z}}$$

$$J(w) = -\sum_{i=1}^{n}[(y_i \times \log(\phi(z_i))) + ((1-y_i) \times (1-\log(\phi(z_i))))]$$

使用以下代码来实现新模型和损失函数：

```
num_outputs = y_train.shape[1]
num_inputs = X_train.shape[1]

learning_rate = 0.001

# 输入图像
x = tf.placeholder(dtype=tf.float32, shape=[None, num_inputs], name="x")
# 输出的类标签
y = tf.placeholder(dtype=tf.float32, shape=[None, num_outputs], name="y")

# 模型参数
w = tf.Variable(tf.zeros([num_inputs,num_outputs]), name="w")
b = tf.Variable(tf.zeros([num_outputs]), name="b")
model = tf.nn.sigmoid(tf.matmul(x, w) + b)

loss = tf.reduce_mean(-tf.reduce_sum(
    (y * tf.log(model)) + ((1 - y) * tf.log(1 - model)), axis=1))
optimizer = tf.train.GradientDescentOptimizer(
    learning_rate=learning_rate).minimize(loss)
```

最后，运行分类模型：

```
num_epochs = 1
with tf.Session() as tfs:
    tf.global_variables_initializer().run()
    for epoch in range(num_epochs):
        tfs.run(optimizer, feed_dict={x: X_train, y: y_train})
        y_pred = tfs.run(tf.argmax(model, 1), feed_dict={x: X_test})
        y_orig = tfs.run(tf.argmax(y, 1), feed_dict={y: y_test})

        preds_check = tf.equal(y_pred, y_orig)
        accuracy_op = tf.reduce_mean(tf.cast(preds_check, tf.float32))
        accuracy_score = tfs.run(accuracy_op)
        print("epoch {0:04d} accuracy={1:.8f}".format(
            epoch, accuracy_score))

        plt.figure(figsize=(14, 4))
        plt.subplot(1, 2, 1)
```

```
plt.scatter(X_test[:, 0], X_test[:, 1], marker='o', c=y_orig)
plt.title('Original')
plt.subplot(1, 2, 2)
plt.scatter(X_test[:, 0], X_test[:, 1], marker='o', c=y_pred)
plt.title('Predicted')
plt.show()
```

上述代码得到了相当好的分类精度（96%），原始和预测数据示意图如图 4-14 所示。

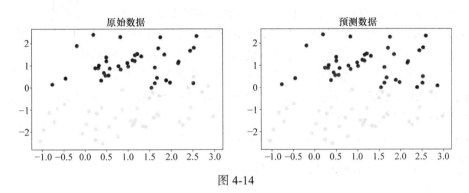

图 4-14

这很简单！！现在让问题变得复杂，即尝试预测两个以上的类别。

4.6 多分类

多分类的一个经典例子是识别手写数字图像。这个例子中的类标签是 {0, 1, 2, 3, 4, 5, 6, 7, 8, 9}。在下面的例子中，将使用 MNIST 数据集。像上一章那样来加载 MNIST 图像，具体代码如下：

```
from tensorflow.examples.tutorials.mnist import input_data
mnist = input_data.read_data_sets(os.path.join(
    datasetslib.datasets_root, 'mnist'), one_hot=True)
```

如果 MNIST 数据集已经加载，则会得到以下输出：

```
Extracting /Users/armando/datasets/mnist/train-images-idx3-ubyte.gz
Extracting /Users/armando/datasets/mnist/train-labels-idx1-ubyte.gz
Extracting /Users/armando/datasets/mnist/t10k-images-idx3-ubyte.gz
Extracting /Users/armando/datasets/mnist/t10k-labels-idx1-ubyte.gz
```

现在来设置一些参数，如下面的代码所示：

```
num_outputs = 10 # 0~9的数字
num_inputs = 784 # 总像素

learning_rate = 0.001
num_epochs = 1
batch_size = 100
num_batches = int(mnist.train.num_examples/batch_size)
```

上述代码中的参数含义分别为：

- num_outputs：由于必须预测图像中数字的类别，因此将输出数量设置为 10。图像对应的是哪个数字，就将输出向量的对应元素置为 1，其他元素全为 0。
- num_inputs：由于输入数字的大小为 28×28 个像素，由于每个像素都是模型的输入，因此共有 784 个输入。
- learning_rate：该参数表示梯度下降优化器算法的学习率，这里将学习率设置为 0.001。
- num_epochs：这里只进行一次迭代，因此将周期数设置为 1。
- batch_size：在实际问题中，数据集可能非常大，为了训练模型，可能需要加载整个数据集，但这是不可行的。因此，需要对数据进行拆分，并将数据随机分配到各个批次中。由于使用 TensorFlow 的内置算法一次可以选择 100 张图像，因此将 batch_size 设置为 100。
- num_batches：该参数设置应从总数据集中选择批次的数量，将其设置为数据集中样本数量除以每个批次的样本数量。

读者可以尝试使用不同值的参数。

下面的代码定义了输入、输出、参数、模型和损失函数：

```
# 输入图像
x = tf.placeholder(dtype=tf.float32, shape=[None, num_inputs], name="x")
# 输出的类标签
y = tf.placeholder(dtype=tf.float32, shape=[None, num_outputs], name="y")

# 模型参数
w = tf.Variable(tf.zeros([784, 10]), name="w")
b = tf.Variable(tf.zeros([10]), name="b")
model = tf.nn.softmax(tf.matmul(x, w) + b)

loss = tf.reduce_mean(-tf.reduce_sum(y * tf.log(model), axis=1))
optimizer = tf.train.GradientDescentOptimizer(
    learning_rate=learning_rate).minimize(loss)
```

这些代码类似于二分类示例，但有一个显著差异：这里使用了 softmax 函数而不是 sigmoid 函数。softmax 用于多分类，而 sigmoid 用于二分类。softmax 函数是 sigmoid 函数的推广形式，它将 n 维向量 z 转换为 n 维向量 $\sigma(z)$，$\sigma(z)$ 各元素的取值范围为 $(0, 1]$，而且这些元素值之和为 1。

现在来运行模型并输出分类精度：

```
with tf.Session() as tfs:
    tf.global_variables_initializer().run()
    for epoch in range(num_epochs):
        for batch in range(num_batches):
            batch_x, batch_y = mnist.train.next_batch(batch_size)
            tfs.run(optimizer, feed_dict={x: batch_x, y: batch_y})
```

```
        predictions_check = tf.equal(tf.argmax(model, 1), tf.argmax(y, 1))
        accuracy_function = tf.reduce_mean(
            tf.cast(predictions_check, tf.float32))
        feed_dict = {x: mnist.test.images, y: mnist.test.labels}
        accuracy_score = tfs.run(accuracy_function, feed_dict)
        print("epoch {0:04d} accuracy={1:.8f}".format(
            epoch, accuracy_score))
```

可以得到如下的分类精度：

```
epoch 0000  accuracy=0.76109999
```

下面尝试通过多次迭代来训练模型，这样可以在每次迭代中学习不同的批次。建立两个函数来帮助完成该任务：

```
def mnist_batch_func(batch_size=100):
    batch_x, batch_y = mnist.train.next_batch(batch_size)
    return [batch_x, batch_y]
```

上面的函数将一批中的样本数作为输入并使用 mnist.train.next_batch( ) 函数返回每个批次中的特征（batch_x）和目标（batch_y）：

```
def tensorflow_classification(num_epochs, num_batches, batch_size,
                              batch_func, optimizer, test_x, test_y):
    accuracy_epochs = np.empty(shape=[num_epochs], dtype=np.float32)
    with tf.Session() as tfs:
        tf.global_variables_initializer().run()
        for epoch in range(num_epochs):
            for batch in range(num_batches):
                batch_x, batch_y = batch_func(batch_size)
                feed_dict = {x: batch_x, y: batch_y}
                tfs.run(optimizer, feed_dict)
            predictions_check = tf.equal(
                tf.argmax(model, 1), tf.argmax(y, 1))
            accuracy_function = tf.reduce_mean(
                tf.cast(predictions_check, tf.float32))
            feed_dict = {x: test_x, y: test_y}
            accuracy_score = tfs.run(accuracy_function, feed_dict)
            accuracy_epochs[epoch] = accuracy_score
            print("epoch {0:04d} accuracy={1:.8f}".format(
                epoch, accuracy_score))

    plt.figure(figsize=(14, 8))
    plt.axis([0, num_epochs, np.min(
        accuracy_epochs), np.max(accuracy_epochs)])
    plt.plot(accuracy_epochs, label='Accuracy Score')
    plt.title('Accuracy over Iterations')
    plt.xlabel('# Epoch')
    plt.ylabel('Accuracy Score')
    plt.legend()
    plt.show()
```

上面的函数使用了一些参数并执行训练迭代，输出每次迭代的分类精度和最终的分类精度。还可以保存 accuracy_epochs 数组中每次迭代的分类精度。然后绘制每次迭代的分类精度。利用上面所设置的参数，使用以下代码来运行 30 次迭代：

```
num_epochs=30
tensorflow_classification(num_epochs=num_epochs,
    num_batches=num_batches,
    batch_size=batch_size,
    batch_func=mnist_batch_func,
    optimizer=optimizer,
    test_x=mnist.test.images,test_y=mnist.test.labels)
```

得到以下分类精度和相应的示意图（见图 4-15）：

```
epoch 0000    accuracy=0.76020002
epoch 0001    accuracy=0.79420000
epoch 0002    accuracy=0.81230003
epoch 0003    accuracy=0.82309997
epoch 0004    accuracy=0.83230001
epoch 0005    accuracy=0.83770001

--- epoch 6 to 24 removed for brevity ---

epoch 0025    accuracy=0.87930000
epoch 0026    accuracy=0.87970001
epoch 0027    accuracy=0.88059998
epoch 0028    accuracy=0.88120002
epoch 0029    accuracy=0.88180000
```

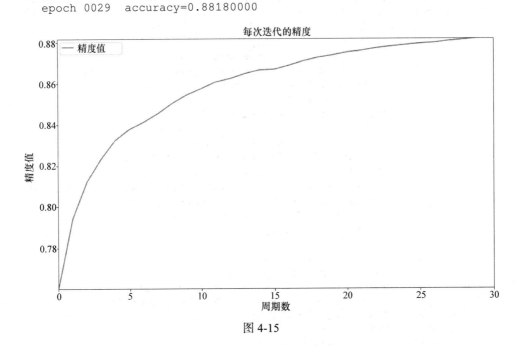

图 4-15

从图中可以看到：在开始时，分类精度提高得非常快，然后提高的速度变慢。稍后将看到如何在 TensorFlow 中使用神经网络来大幅提高分类精度。

4.7 总结

本章介绍了 TensorFlow 中的经典机器学习算法。本章的第一部分介绍了回归模型，并给出如何训练具有一个或多个特征的线性回归模型，以及如何使用 TensorFlow 来编写线性回归的代码。正则化其实就是增加一个惩罚项，以便学习参数时，模型不会过度拟和训练数据，本章使用 TensorFlow 实现了 Lasso 正则化、岭正则化和弹性网络正则化。在接下来的章节中会介绍 TensorFlow 内置的正则化方法。

本章后面的内容介绍了监督机器学习中的分类问题，讨论了二类和多分类的模型函数、平滑函数和损失函数。本章介绍了 Logistic 回归，因为这是实现分类的最简单的方法。对于二分类，使用 sigmoid 函数，对于多分类，使用 softmax 函数来平滑线性模型的值，以输出在各个类别中的概率。

用 TensorFlow 实现了 Logistic 模型和相应的损失函数，并且训练了二分类和多分类的模型。虽然在本章中介绍了经典的机器学习方法，并使用 TensorFlow 来实现它们，但是当用神经网络和深度神经网络来解决机器学习问题时，就能充分体现 TensorFlow 的优越性。本书将在神经网络的相关章节中研究这些先进的方法。

鼓励读者阅读以下书籍，以便能了解有关回归和分类的更多细节：
Sebastian Raschka, *Python Machine Learning, 2nd Edition*. Packt Publishing, 2017
Trevor Hastie，Robert Tibshirani，Jerome Friedman，*The Elements of Statistical Learning*. Second Edition Springer，2013

第 5 章
基于 TensorFlow 和 Keras 的神经网络和多层感知机

神经网络是一种建模技术，该技术是受大脑的结构和功能的启发而提出的。就像大脑含有数百万被称为神经元的微小互连单元一样，现在的神经网络也由数以百万计的小型互联计算单元组成。由于神经网络的计算单元仅存在于数字世界中，为了与大脑的物理神经元进行区分，神经网络的计算单元也被称为人工神经元。类似地，神经网络（NN）也被称为人工神经网络（ANN）。

本章将介绍以下主题：
- 感知机。
- 多层感知机。
- 用于图像分类的多层感知机：
  - 通过 TensorFlow 构建用于 MNIST 分类的多层感知机；
  - 通过 Keras 构建用于 MNIST 分类的多层感知机；
  - 通过 TFLearn 构建用于 MNIST 分类的多层感知机。
- 用于时间序列回归的多层感知机。

5.1 感知机

首先介绍一下神经网络最基本的模块，即**感知机**，也称为**人工神经元**。感知机的概念起源于 Frank Rosenblatt 在 1962 年所做的工作。

> 可以阅读以下资料来研究神经网络的起源：
> Frank Rosenblatt，*Principles of Neurodynamics: Perceptrons and the Theory of Brain Mechanisms*. Spartan Books, 1962.

简而言之，感知机模仿的是生物神经元，它需要一个或多个输入并将它们组合以产生输出。

如图 5-1 所示，感知机接收三个输入并将其相加来得到输出 y。

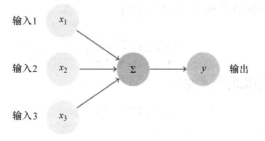

图 5-1

这个感知机太简单了，没有任何实际用途。不过，通过增加权重、偏差和激活函数，可以让其增强。为每个输入求加权和 $\sum w_i x_i$，如果加权和小于阈值，则输出为 0，否则输出为 1：

$$y = \begin{cases} 0 & \text{如果} \sum w_i x_i < \text{阈值} \\ 0 & \text{如果} \sum w_i x_i \geq \text{阈值} \end{cases}$$

阈值被称为偏差（bias），将偏差移到等式左边，并用 b 表示，用 w 和 x 的向量内积表示 $\sum w_i x_i$。现在，感知机可表示为：

$$y = \begin{cases} 1 & \text{如果} \sum wx + b < 0 \\ 1 & \text{如果} \sum wx + b \geq 0 \end{cases}$$

现在，感知机看起来像图 5-2 所示的图像。

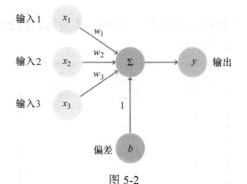

图 5-2

目前为止，神经元是一个线性函数。为了使神经元产生一个非线性决策边界，可通过非线性的**激活**（activation）函数来计算求和后的输出。有许多流行的激活函数可以使用：

- ReLU：**整流线性单元**（Rectified Linear Unit），它将值平滑到 $(0, x)$ 之间：

$$\text{ReLU}(x) = \max(0, x)$$

- sigmoid: sigmoid 将值平滑到 $(0, 1)$ 之间：

$$\text{sigmoid}(x) = \frac{1}{1 + e^{-x}} = \frac{e^x}{1 + e^x}$$

- tanh：**双曲正切**将值平滑到（-1，1）之间：

$$\tanh(x) = \frac{e^x - e^{-x}}{e^x + e^{-x}}$$

利用激活函数，感知机的公式变为：

$$y = \varphi(wx + b)$$

式中，$\varphi(\cdot)$ 是激活函数。

图 5-3 为神经元的示意图。

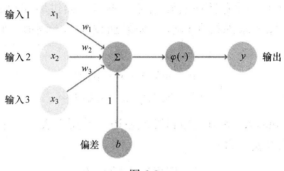

图 5-3

5.2 多层感知机

通过确定的结构将人工神经元连接在一起时，就是所谓的神经网络。图 5-4 为一个最简单的神经网络，它只有一个神经元。

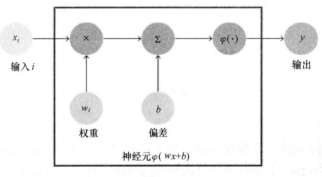

图 5-4

把神经元串联起来，使得上一层的输出成为下一层的输入，这种神经网络被称为**前馈神经网络（FFNN）**。这些 FFNN 由连接在一起的神经元层组成，因此被称为**多层感知机（MLP）或深度神经网络（DNN）**。

例如，图 5-5 中描述的 MLP 将三个特征作为输入，有两个隐藏层，每个隐藏层各有五个神经元，还有一个输出 y，上一层的神经元与下一层神经元之间是完全连接的，这样的层也被称为致密层或仿射层，并且这样的模型也被称为顺序模型（sequential model）。

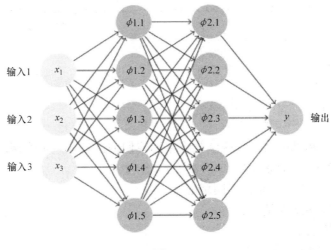

图 5-5

下面介绍在 TensorFlow 中构建简单的神经网络（MLP 或 DNN），这里会使用之前介绍过的数据集。

可以按照 Jupyter 笔记本 ch-05_MLP 中的代码进行操作。

5.3 用于图像分类的多层感知机

使用 TensorFlow、Keras 和 TFLearn 等不同库构建用于图像分类的 MLP 网络。所使用的数据集为 MNIST。

MNIST 数据集包含了从 0 到 9 的手写数字（每幅图像的大小为 28×28 像素），以及对应的标签，训练集大小为 60K，测试集大小为 10K。在本书的 TensorFlow 示例和教程中，会广泛使用 MNIST 数据集。

MNIST 数据集和相关文档可从以下链接获得：http://yann.lecun.com/exdb/mnist/。

现在从纯粹的 TensorFlow 方法开始。

5.3.1 通过 TensorFlow 构建用于 MNIST 分类的多层感知机

首先，加载 MNIST 数据集，并使用以下代码定义训练特征和测试特征以及类标签：

```
from tensorflow.examples.tutorials.mnist import input_data
mnist_home = os.path.join(datasetslib.datasets_root, 'mnist')
mnist = input_data.read_data_sets(mnist_home, one_hot=True)

X_train = mnist.train.images
X_test = mnist.test.images
Y_train = mnist.train.labels
Y_test = mnist.test.labels

num_outputs = 10  # 0~9 的数字
num_inputs = 784  # 总像素
```

还需要创建三个辅助函数,这些函数将帮助我们创建只有一个隐藏层的简单 MLP,然后再创建较大的 MLP,它有多个层,每层中有多个神经元。

mlp( ) 函数使用以下逻辑来构建网络层。

1)mlp( ) 函数需要五个输入:
- x 是输入的特征张量;
- num_inputs 是输入特征的数量;
- num_outputs 是输出目标的数量;
- num_layers 是所需的隐藏层数;
- num_neurons 是包含每个网层的神经元数量的列表。

2)将权重和偏差设置为空:

```
w=[]
b=[]
```

3)以隐藏层的数量作为循环次数,以创建权重和偏差张量,并将张量加到各自的列表中:

- 张量被分别命名为 w_<layer_num> 和 b_<layer_num>,命名张量有助于调试并定位问题代码。
- 使用正态分布 tf.random_normal( ) 对张量进行初始化。
- 权重张量的第一维是前一层输入的数量,对于第一个隐藏层,第一个维度是 num_inputs。权重张量的第二维是当前网络层的神经元数量。
- 偏差都是一维张量,维数等于当前层中神经元的数量。

```
for i in range(num_layers):
    # 权重
    w.append(tf.Variable(tf.random_normal(
        [num_inputs if i == 0 else num_neurons[i - 1],
         num_neurons[i]]),
        name="w_{0:04d}".format(i)
        ))
    # 偏差
    b.append(tf.Variable(tf.random_normal(
        [num_neurons[i]]),
        name="b_{0:04d}".format(i)
        ))
```

4）为最后一个隐藏层创建权重和偏差。在这种情况下，权重张量的维数等于最后隐藏层中神经元的数量和输出目标的数量。偏差是单个维度的张量，其大小为输出特征的数量：

```
w.append(tf.Variable(tf.random_normal(
    [num_neurons[num_layers - 1] if num_layers > 0 else num_inputs,
     num_outputs]), name="w_out"))
b.append(tf.Variable(tf.random_normal([num_outputs]),
    name="b_out"))
```

5）现在开始定义网络层。首先，很明显，x 是第一个输入层：

```
# x 是输入层
layer = x
```

6）在循环中添加隐藏层。每个隐藏层表示通过激活函数 tf.nn.relu( ) 把线性函数 tf.matmul(layer, w [i]) + b [i] 转变成非线性函数：

```
# 添加隐藏层
for i in range(num_layers):
    layer = tf.nn.relu(tf.matmul(layer, w[i]) + b[i])
```

7）添加输出层。输出层和隐藏层之间的一个区别在于输出层中没有激活函数：

```
layer = tf.matmul(layer, w[num_layers]) + b[num_layers]
```

8）返回包含 MLP 的网络层对象：

```
return layer
```

实现整个 MLP 的完整代码如下：

```
def mlp(x, num_inputs, num_outputs, num_layers, num_neurons):
    w = []
    b = []
    for i in range(num_layers):
        # 权重
        w.append(tf.Variable(tf.random_normal(
            [num_inputs if i == 0 else num_neurons[i - 1],
             num_neurons[i]]),
            name="w_{0:04d}".format(i)
        ))
        # 偏差
        b.append(tf.Variable(tf.random_normal(
            [num_neurons[i]]),
            name="b_{0:04d}".format(i)
        ))
    w.append(tf.Variable(tf.random_normal(
        [num_neurons[num_layers - 1] if num_layers > 0 else num_inputs,
         num_outputs]), name="w_out"))
    b.append(tf.Variable(tf.random_normal([num_outputs]), name="b_out"))

    # x 是输入层
    layer = x
    # 添加隐藏层
    for i in range(num_layers):
        layer = tf.nn.relu(tf.matmul(layer, w[i]) + b[i])
```

```
# 添加输出层
layer = tf.matmul(layer, w[num_layers]) + b[num_layers]

return layer
```

辅助函数 mnist_batch_func( ) 封装了针对 MNIST 数据集的批处理函数，以便能获取下一批图像：

```
def mnist_batch_func(batch_size=100):
    X_batch, Y_batch = mnist.train.next_batch(batch_size)
    return [X_batch, Y_batch]
```

这个函数的功能不言自明。TensorFlow 为 MNIST 数据集提供此函数，但是，对于其他数据集，可能需要编写独立的批处理函数。

辅助函数 tensorflow_classification( ) 用于训练和评估模型。

1）tensorflow_classification( ) 函数有如下的参数：

- n_epochs 是训练要运行的循环次数；
- n_batches 是训练的每个循环中对随机采样的样本批次数；
- batch_size 是每批中的样本数量；
- batch_func 是获取 batch_size 并返回 X 和 Y 样本批次的函数；
- model 是具有神经元的神经网络或网络层；
- optimizer 是指 TensorFlow 定义的优化函数；
- loss 是指优化器进行参数优化的损失函数；
- accuracy_function 是计算分类精度的函数；
- X_test 和 Y_test 是测试的数据集。

2）启动 TensorFlow 会话来进行训练：

```
with tf.Session() as tfs:
    tf.global_variables_initializer().run()
```

3）进行 n_epoch 次训练：

```
for epoch in range(n_epochs):
```

4）在每个循环中，取 n_batches 个样本集来训练模型，计算每个批次的损失函数值，计算每次循环的平均损失函数值：

```
epoch_loss = 0.0
        for batch in range(n_batches):
            X_batch, Y_batch = batch_func(batch_size)
            feed_dict = {x: X_batch, y: Y_batch}
            _, batch_loss = tfs.run([optimizer, loss], feed_dict)
            epoch_loss += batch_loss
        average_loss = epoch_loss / n_batches
        print("epoch: {0:04d} loss = {1:0.6f}".format(
            epoch, average_loss))
```

5）当所有迭代循环完成后，输出由 accuracy_function 函数计算的分类精度：

```
feed_dict = {x: X_test, y: Y_test}
accuracy_score = tfs.run(accuracy_function,
                    feed_dict=feed_dict)
print("accuracy={0:.8f}".format(accuracy_score))
```

tensorflow_classification( ) 函数的完整代码如下所示：

```
def tensorflow_classification(n_epochs, n_batches,
                    batch_size, batch_func,
                    model, optimizer, loss, accuracy_function,
                    X_test, Y_test):
    with tf.Session() as tfs:
        tfs.run(tf.global_variables_initializer())
        for epoch in range(n_epochs):
            epoch_loss = 0.0
            for batch in range(n_batches):
                X_batch, Y_batch = batch_func(batch_size)
                feed_dict = {x: X_batch, y: Y_batch}
                _, batch_loss = tfs.run([optimizer, loss], feed_dict)
                epoch_loss += batch_loss
            average_loss = epoch_loss / n_batches
            print("epoch: {0:04d} loss = {1:0.6f}".format(
                epoch, average_loss))
        feed_dict = {x: X_test, y: Y_test}
        accuracy_score = tfs.run(accuracy_function, feed_dict=feed_dict)
        print("accuracy={0:.8f}".format(accuracy_score))
```

现在来定义输入和输出的占位符 x 和 y，以及其他超参数：

```
# 输入图像
x = tf.placeholder(dtype=tf.float32, name="x",
                    shape=[None, num_inputs])
# 目标输出
y = tf.placeholder(dtype=tf.float32, name="y",
                    shape=[None, num_outputs])
num_layers = 0
num_neurons = []
learning_rate = 0.01
n_epochs = 50
batch_size = 100
n_batches = int(mnist.train.num_examples/batch_size)
```

各个参数的含义如下：

- num_layers 是隐藏层的数量。最开始介绍的神经网络没有隐藏层，只有输入层和输出层。
- num_neurons 是空列表，因为没有隐藏层。
- learning_rate 是 0.01，一个随机选择的小数字。
- num_epochs 表示 50 次迭代，以了解将输入连接到输出的唯一神经元的参数。
- batch_size 为 100，这个参数可能取不同值。每个批次较大不一定会带来更好的效果。为了给神经网络找到最佳的 batch_size 值，可能需要研究不同的 batch_size 取值。

- **n_batches**：批次数量。它大致等于训练样本数除以每个批次的样本数。

现在把所有东西放在一起，使用到目前为止得到的变量来定义网络、损失函数、优化器函数和分类精度函数。

```
model = mlp(x=x,
            num_inputs=num_inputs,
            num_outputs=num_outputs,
            num_layers=num_layers,
            num_neurons=num_neurons)

loss = tf.reduce_mean(
    tf.nn.softmax_cross_entropy_with_logits(logits=model, labels=y))
optimizer = tf.train.GradientDescentOptimizer(
    learning_rate=learning_rate).minimize(loss)

predictions_check = tf.equal(tf.argmax(model, 1), tf.argmax(y, 1))
accuracy_function = tf.reduce_mean(tf.cast(predictions_check, tf.float32))
```

在上面的代码中，使用一个新的张量函数来定义损失函数：

```
tf.nn.softmax_cross_entropy_with_logits(logits=model, labels=y)
```

当使用 softmax_cross_entropy_with_logits( ) 函数时，请确保输出未缩放并且尚未传递给 softmax 激活函数。softmax_cross_entropy_with_logits 函数在内部使用 softmax 来缩放输出。

该函数计算模型（估计值 y）和 y 的实际值之间的 softmax 熵。当输出属于某一个类别，并且不超过一个类时，可使用熵函数。正如在例子中提到的，每幅图像只有一个数字。

关于熵函数的更多信息参考：https://www.tensorflow.org/api_docs/python/tf/nn/softmax_cross_entropy_with_logits。

接下来可运行 tensorflow_classification 函数来训练和评估模型：

```
tensorflow_classification(n_epochs=n_epochs,
    n_batches=n_batches,
    batch_size=batch_size,
    batch_func=mnist_batch_func,
    model = model,
    optimizer = optimizer,
    loss = loss,
    accuracy_function = accuracy_function,
    X_test = mnist.test.images,
    Y_test = mnist.test.labels
    )
```

在运行过程中会得到以下输出：

```
epoch: 0000   loss = 8.364567
epoch: 0001   loss = 4.347608
```

```
epoch: 0002    loss = 3.085622
epoch: 0003    loss = 2.468341
epoch: 0004    loss = 2.099220
epoch: 0005    loss = 1.853206

--- Epoch 06 to 45 output removed for brevity ---

epoch: 0046    loss = 0.684285
epoch: 0047    loss = 0.678972
epoch: 0048    loss = 0.673685
epoch: 0049    loss = 0.668717
accuracy=0.85720009
```

看到单个神经网络在 50 次迭代中缓慢地将损失函数的值从 8.3 降低到 0.66，最终得到的分类精度为 85% 左右。这个例子的分类精度不太好，因为这只是使用 TensorFlow 进行 MLP 分类的演示。

使用更多的网络层和神经元来执行相同的代码，可以获得表 5-1 所示分类精度。

表 5-1

层数	每个隐藏层中的神经元数量	分类精度
0	0	0.857
1	8	0.616
2	256	0.936

因此，通过增加两个隐藏层，每个隐藏层有 256 个神经元，则可将分类精度提高到 0.936，鼓励大家尝试使用具有不同变量值的代码来观察损失和分类精度的变化情况。

5.3.2 通过 Keras 构建用于 MNIST 分类的多层感知机

现在用 Keras（一个 TensorFlow 的高级库）来建立同样的 MLP 神经网络。让所有参数与前面示例的参数相同，例如，隐藏层的激活函数仍采用 ReLU 函数。

1）从 Keras 中导入所需的模块：

```
import keras
from keras.models import Sequential
from keras.layers import Dense
from keras.optimizers import SGD
```

2）定义超参数（假设数据集已经加载到变量 X_train、Y_train、X_test 和 Y_test 中）：

```
num_layers = 2
num_neurons = []
for i in range(num_layers):
    num_neurons.append(256)
learning_rate = 0.01
n_epochs = 50
batch_size = 100
```

3）创建一个顺序模型：

```
model = Sequential()
```

4）添加第一个隐藏层。在第一个隐藏层中必须指定输入张量的形状：

```
model.add(Dense(units=num_neurons[0], activation='relu',
    input_shape=(num_inputs,)))
```

5）添加第二层：

```
model.add(Dense(units=num_neurons[1], activation='relu'))
```

6）添加具有激活函数 softmax 的输出层：

```
model.add(Dense(units=num_outputs, activation='softmax'))
```

7）输出模型的详细信息：

```
model.summary()
```

可以得到以下结果：

```
Layer (type)                 Output Shape              Param #
=================================================================
dense_1 (Dense)              (None, 256)               200960
_____
dense_2 (Dense)              (None, 256)               65792
_____
dense_3 (Dense)              (None, 10)                2570
_____
Total params: 269,322
Trainable params: 269,322
Non-trainable params: 0
```

8）使用 SGD 优化器编译模型：

```
model.compile(loss='categorical_crossentropy',
    optimizer=SGD(lr=learning_rate),
    metrics=['accuracy'])
```

9）训练模型

```
model.fit(X_train, Y_train,
    batch_size=batch_size,
    epochs=n_epochs)
```

在模型的训练过程中可以观察到每次训练迭代的损失函数值和分类精度：

```
Epoch 1/50
55000/55000 [======================] - 4s - loss: 1.1055 - acc: 0.7413
Epoch 2/50
55000/55000 [======================] - 3s - loss: 0.4396 - acc: 0.8833
Epoch 3/50
55000/55000 [======================] - 3s - loss: 0.3523 - acc: 0.9010
Epoch 4/50
55000/55000 [======================] - 3s - loss: 0.3129 - acc: 0.9112
Epoch 5/50
55000/55000 [======================] - 3s - loss: 0.2871 - acc: 0.9181
```

```
--- Epoch 6 to 45 output removed for brevity ---
Epoch 46/50
55000/55000 [==============================] - 4s - loss: 0.0689 - acc: 0.9814
Epoch 47/50
55000/55000 [==============================] - 4s - loss: 0.0672 - acc: 0.9819
Epoch 48/50
55000/55000 [==============================] - 4s - loss: 0.0658 - acc: 0.9822
Epoch 49/50
55000/55000 [==============================] - 4s - loss: 0.0643 - acc: 0.9829
Epoch 50/50
55000/55000 [==============================] - 4s - loss: 0.0627 - acc: 0.9829
```

10）评估模型并输出损失函数值和分类精度：

```
score = model.evaluate(X_test, Y_test)
print('\n Test loss:', score[0])
print('Test accuracy:', score[1])
```

得到如下输出结果：

```
Test loss: 0.089410082236
Test accuracy: 0.9727
```

在 ch-05_MLP 中提供了使用 Keras 进行 MNIST 分类的完整 MLP 代码。

5.3.3 通过 TFLearn 构建用于 MNIST 分类的多层感知机

下面介绍如何使用 TFLearn（另一个 TensorFlow 的高级库）来实现相同的 MLP。

1）导入 TFLearn 库：

```
import tflearn
```

2）定义超参数（假设数据集已经加载到变量 X_train、Y_train、X_test 和 Y_test 中）：

```
num_layers = 2
num_neurons = []
for i in range(num_layers):
num_neurons.append(256)

learning_rate = 0.01
n_epochs = 50
batch_size = 100
```

3）构建输入层、两个隐藏层和输出层（与 TensorFlow 和 Keras 中的示例架构相同）：

```
# 构建DNN
input_layer = tflearn.input_data(shape=[None, num_inputs])
dense1 = tflearn.fully_connected(input_layer, num_neurons[0],
    activation='relu')
dense2 = tflearn.fully_connected(dense1, num_neurons[1],
    activation='relu')
softmax = tflearn.fully_connected(dense2, num_outputs,
    activation='softmax')
```

4）为最后一步建立的 DNN（在变量 softmax 中）指定优化器、神经网络和 MLP 模型（在 TFLearn 中称为 DNN）：

```
optimizer = tflearn.SGD(learning_rate=learning_rate)
net = tflearn.regression(softmax, optimizer=optimizer,
                    metric=tflearn.metrics.Accuracy(),
                    loss='categorical_crossentropy')
model = tflearn.DNN(net)
```

5）训练模型：

```
model.fit(X_train, Y_train, n_epoch=n_epochs,
         batch_size=batch_size,
         show_metric=True, run_id="dense_model")
```

训练完成后，会得到如下结果：

```
Training Step: 27499  | total loss: 0.11236 | time: 5.853s
| SGD | epoch: 050 | loss: 0.11236 - acc: 0.9687 -- iter: 54900/55000
Training Step: 27500  | total loss: 0.11836 | time: 5.863s
| SGD | epoch: 050 | loss: 0.11836 - acc: 0.9658 -- iter: 55000/55000
--
```

6）评估模型并给出分类精度：

```
score = model.evaluate(X_test, Y_test)
print('Test accuracy:', score[0])
```

得到以下输出结果：

```
Test accuracy: 0.9637
```

使用 TFLearn 也可以获得非常高的分类精度。

ch-05_MLP 中提供了使用 TFLearn 进行 MNIST 分类的完整 MLP 代码。

5.3.4 多层感知机与 TensorFlow、Keras 和 TFLearn 的总结

前面介绍如何使用 TensorFLow 及其高级库构建简单的 MLP 网络。纯 TensorFlow 所构建的神经网络模型的分类精度为 0.93~0.94，Keras 为 0.96~0.98，TFLearn 为 0.96~0.97。尽管所有的代码示例都是基于 TensorFlow 的，但同一架构和参数的精度差异会让人得出这样一个事实：尽管对一些重要的超参数进行了初始化，但高级库和 TensorFlow 抽象了许多其他的超参数，我们无法修改这些参数的默认值。

还可以看到，与 Keras 和 TFLearn 相比，TensorFlow 的代码比较冗长。因此，高级库使得构建和训练神经网络模型变得更容易。

5.4 用于时间序列回归的多层感知机

前面已经介绍了图像数据分类的例子，现在来看一下基于时间序列数据的回归例子。这里将使用 MLP 来针对国际航空公司乘客数据集构建一个较小的单变量时间序列模型。该数据集包含多年积累的乘客数，数据集可在以下链接中找到：

- https://www.kaggle.com/andreazzini/international-airline-passengers/data

- https://datamarket.com/data/set/22u3/international-airline-passengers-monthly-totals-in-thousands-jan-49-dec-60

开始准备数据集。

1) 首先，使用以下代码加载数据集：

```
filename = os.path.join(datasetslib.datasets_root,
                        'ts-data',
                        'international-airline-passengers-cleaned.csv')
dataframe = pd.read_csv(filename,usecols=[1],header=0)
dataset = dataframe.values
dataset = dataset.astype('float32')
```

2) 使用 datasetslib 所提供的工具将数据集分成测试集和训练集。对于时间序列数据集，有一个单独的函数，因为对于时间序列回归，需要维护好观察的顺序。使用 67% 的数据作为训练数据集，33% 的数据作为测试数据集。也可以尝试使用不同比例。

```
train,test=dsu.train_test_split(dataset,train_size=0.67)
```

3) 对于时间序列回归，需要把数据集转换成监督数据集。在这个例子中，使用两个时间步长的滞后。将 n_x 设置为 2，并且使用 mvts_to_xy() 函数返回训练数据集和测试数据集的输入和输出（X 和 Y），使得 X 有两列时间值 {t-1, t}，而 Y 只有一列时间值 {t + 1}。学习算法假设 t + 1 时刻的值可以通过时间 {t-1, t, t + 1} 之间的关系来学习得到。

```
# 重塑为 X=t-1,t 和 Y=t+1
n_x=2
n_y=1
X_train, Y_train, X_test, Y_test = tsd.mvts_to_xy(train,
                                    test,n_x=n_x,n_y=n_y)
```

 有关将时间序列数据集转换为监督学习问题的更多信息可以在以下链接中找到：http:// machinelearningmastery.com/convert-time-series-supervised-learning-problem-python/。

现在在训练数据集上建立和训练模型。

1) 导入所需的 Keras 模块：

```
from keras.models import Sequential
from keras.layers import Dense
from keras.optimizers import SGD
```

2) 设置构建模型所需的超参数：

```
num_layers = 2
num_neurons = [8,8]
n_epochs = 50
batch_size = 2
```

请注意，由于数据集非常小，因此批处理大小为 2。由于问题的规模很小，因此使用所建立的 MLP 只有两层，每层只有 8 个神经元。

3）建立、编译和训练模型：

```
model = Sequential()
model.add(Dense(num_neurons[0], activation='relu',
    input_shape=(n_x,)))
model.add(Dense(num_neurons[1], activation='relu'))
model.add(Dense(units=1))
model.summary()

model.compile(loss='mse', optimizer='adam')

model.fit(X_train, Y_train,
    batch_size=batch_size,
    epochs=n_epochs)
```

 请注意，这里用 Adam 优化器代替 SGD。读者可以尝试用 TensorFlow 和 Keras 中的其他优化器。

4）评估模型并输出均方误差（MSE）和均方根误差（RMSE）：

```
score = model.evaluate(X_test, Y_test)
print('\nTest mse:', score)
print('Test rmse:', math.sqrt(score))
```

得到以下输出结果：

```
Test mse: 5619.24934188
Test rmse: 74.96165247566114
```

5）在测试数据集和训练数据集上使用训练好的模型来预测并绘制这些值：

```
# 设置预测值
Y_train_pred = model.predict(X_train)
Y_test_pred = model.predict(X_test)

# 改变训练预测值用于绘图
Y_train_pred_plot = np.empty_like(dataset)
Y_train_pred_plot[:, :] = np.nan
Y_train_pred_plot[n_x-1:len(Y_train_pred)+n_x-1, :] = Y_train_pred

# 改变测试预测值用于绘图
Y_test_pred_plot = np.empty_like(dataset)
Y_test_pred_plot[:, :] = np.nan
Y_test_pred_plot[len(Y_train_pred)+(n_x*2)-1:len(dataset)-1, :] = \
    Y_test_pred

# 绘制基线和预测值
plt.plot(dataset,label='Original Data')
plt.plot(Y_train_pred_plot,label='Y_train_pred')
plt.plot(Y_test_pred_plot,label='Y_test_pred')
plt.legend()
plt.show()
```

图 5-6 展示了原来的时间序列值和预测的时间序列值。

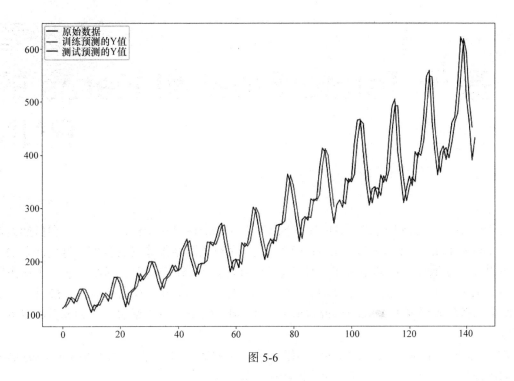

图 5-6

正如所看到的,这是一个相当不错的估计。但在现实生活中,数据基本上都是多变量,其情形会很复杂。因此,在下面的章节中将介绍基于时间序列数据的循环神经网络(RNN)模型。

5.5 总结

本章介绍了多层感知机(MLP),解释了如何构建、训练用于解决分类和回归问题的 MLP 模型,并使用纯 TensorFlow、Keras 和 TFLearn 构建了 MLP 模型。其中,对于分类问题,使用图像数据;而对于回归问题,使用了时间序列数据。

构建和训练 MLP 网络模型的技术对于其他类型的数据(如数字或文本)都是相同的。然而,对于图像数据集,CNN 体系结构已被证明是最好的;对于序列数据集(如时间序列和文本),RNN 模型已被证明是最好的。

虽然本章仅使用简单的数据集示例来演示 MLP 体系结构,但在后面的章节中,将使用一些大型和高级数据集来介绍 CNN 和 RNN 的体系结构。

第 6 章 基于 TensorFlow 和 Keras 的 RNN

对于时序数据问题 [例如，**时间序列预测**（Time Series Forecasting，TSF）和**自然语言处理**（Natural Language Processing, NLP）]，上下文对预测结果非常有价值。这些问题的上下文可以通过获取整个序列来确定，而不仅仅是最后一个数据样本。因此，前一个输出需要成为当前输入的一部分，当重复时，最后一个输出需要成为所有先前输入和最后一个输入的结果。**循环神经网络**（Recurrent Neural Network，RNN）可以解决与序列相关的机器学习问题。

RNN 是处理序列数据的专用神经网络架构。序列数据可以是一段时间内的观察顺序，如时间序列数据，或字符、单词和句子的顺序，如文本数据。

标准神经网络的一个假设是输入数据的排列方式是一个输入不依赖另一个输入，然而，对于时序数据和文本数据而言，这种假设并不适用，因为序列后面出现的值常常受到之前出现的值的影响。

为了实现这一点，RNN 在标准神经网络的基础上按下列方式进行了扩展：
- RNN 通过在计算图中增加循环，将某一网络层的输出作为同一层（或上一层）的输入。
- RNN 添加了记忆单元，用于存储先前的输入和输出，这些会在当前计算中用到。

本章将介绍以下与 RNN 相关的主题：
- 简单 RNN；
- RNN 改进版本；
- LSTM 网络；
- GRU 网络；
- 基于 TensorFlow 的 RNN；
- 基于 Keras 的 RNN；
- 基于 Keras 的 RNN 用于 MNIST 数据。

接下来的两章将介绍在 TensorFlow 和 Keras 中为时间序列和文本（NLP）数据构建 RNN 模型的例子。

6.1 简单 RNN

图 6-1 是一个带有循环的简单神经网络。

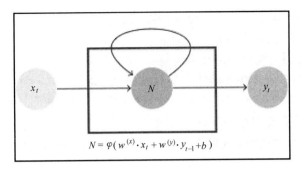

图 6-1

在图 6-1 中，神经网络 N 采用输入 x_t 来产生输出 y_t。由于具有循环结构，在下一个时间点 $t+1$，将输入 y_t 与输入 x_{t+1} 一起产生输出 y_{t+1}。在数学上，将其表示为：

$$y_t = \varphi\left(w^{(x)} \cdot x_t + w^{(y)} \cdot y_{t-1} + b\right)$$

图 6-2 为按时间步长展开循环的 RNN 结构。

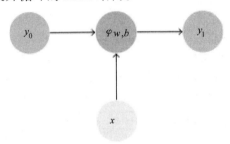

图 6-2

随着时间的演变，此循环在时间步长 5 处展开的示意图如图 6-3 所示。

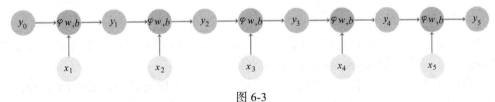

图 6-3

在每个时间步长中，使用相同的学习函数 $\varphi(\cdot)$，以及相同的参数 w 和 b。

输出 y 并不总是在每个时间步长产生。相反，输出 h 是在每个时间步长产生，另一个激活函数应用于这个输出 h 以产生输出 y。这样 RNN 的公式可表示为：

$$h_t = \varphi\left(w^{(hx)} \cdot x_t + w^{(hh)} \cdot h_{t-1} + b^{(h)}\right)$$
$$y_t = \varphi\left(w^{(yh)} \cdot h_t + b^{(y)}\right)$$

式中

- $w^{(hx)}$ 是输入 x 的权向量，该输入连接到隐藏层；
- $w^{(hh)}$ 是上一时间步长的 h 值的权重向量；
- $w^{(yh)}$ 是连接隐藏层与输出层的权重向量；
- 用于 h_t 的函数通常是非线性函数，如 tanh 或 ReLU。

在 RNN 中，每个时间步长使用相同的参数（$w^{(hx)}$，$w^{(hh)}$，$w^{(yh)}$，$b^{(h)}$，$b^{(y)}$）。这样会大大减少为得到序列模型而需要学习的参数的数量。

RNN 在时间步长 t_5 处按图 6-4 的方式展开，假设输出 y 仅在时间步长 t_5 处产生。

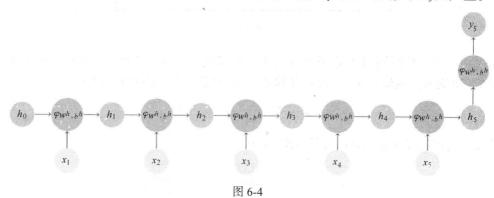

图 6-4

简单的 RNN 在 1990 年由 Elman 引入，因此也被称为 Elman 网络。然而，简单的 RNN 并不能满足今天的处理需求，因此下一节将介绍 RNN 的改进版本。

可阅读 Elman 的原始论文来了解 RNN 架构的起源：
J. L. Elman, Finding Structure in Time, Cogn. Sci., vol. 14, no. 2, pp. 179–211, 1990.

6.2 RNN 改进版本

RNN 架构在很多方面得到了扩展，以适应某些问题的额外需求，并克服了简单 RNN 模型的缺点。本节将介绍 RNN 架构的一些重要的改进。

- 当输出取决于序列的前一个和后一个元素时，可使用**双向 RNN（BRNN）**。BRNN 通过堆叠两个 RNN（称为前向层和后向层）来实现，并且输出都是 RNN 隐藏状态的结果。在前向层中，记忆状态 h 从时间步长 t 流向时间步长 $t+1$，而在后向层中记忆状态从时间步长 t 流向时间步长 $t-1$。两个层在时间步长 t 处采用相同的输入 x_t，但在时间步长 t 处共同产生输出。

- **深度双向 RNN（DBRNN）**进一步扩展了 BRNN，添加了多个网络层。BRNN 在整个时间中都有隐藏的层或跨越时间维度的单元。但是，通过堆叠 BRNN，可以让 DBRNN 具有层次结构。其中一个显著的区别是在 BRNN 中，为同一层的每个单元格使用相同的参数，但在 DBRNN 中每个堆叠层使用不同的参数。

- 为了计算隐藏状态，**长短期记忆（LSTM）网络**通过多个非线性函数来代替一个简单的非线性函数，从而扩展了 RNN。LSTM 网络由称为单元格的黑框组成，采用三个输入：时间 $t–1$ 的工作内存（h_{t-1}）、当前的输入（x_t）和时间 $t–1$ 的长期记忆（c_{t-1}），并产生两个输出：更新的工作记忆（h_t）和长期工作记忆（c_t）。单元使用称为门（gate）的函数来做出从记忆中保存内容或删除内容的决定。下面的章节会详细介绍 LSTM 网络。

阅读以下有关 LSTM 网络的研究文章，以获取更多关于 LSTM 网络起源的信息：
S. Hochreiter and J. Schmidhuber, Long Short-Term Memory, Neural Comput., vol. 9, no. 8, pp. 1735–1780, 1997. http://www.bioinf.jku.at/ publications/older/2604.pdf.

- **门控循环单元（GRU）网络**是 LSTM 网络的简化版本，通过较简单地更新门来结合遗忘和输入门的功能，还将隐藏状态和单元状态组合成单一状态。因此，与 LSTM 网络相比，GRU 网络在计算上更有效。下面将详细介绍 GRU 网络。

读者若想了解 GRU 网络的更多细节，可阅读以下研究论文：
K. Cho, B. van Merrienboer, C. Gulcehre, D. Bahdanau, F. Bougares, H. Schwenk, and Y. Bengio, Learning Phrase Representations using RNN Encoder-Decoder for Statistical Machine Translation, 2014. https:// arxiv.org/abs/1406.1078.
J. Chung, C. Gulcehre, K. Cho, and Y. Bengio, Empirical Evaluation of Gated Recurrent Neural Networks on Sequence Modeling, pp. 1–9, 2014. https://arxiv.org/abs/1412.3555.

- **seq2seq 模型**将编码器 - 解码器架构与 RNN 架构相结合。在 seq2seq 体系结构中，对数据序列（例如文本数据或时间序列数据）进行训练，然后使用该模型生成输出序列。例如，在英文文本上训练模型，然后通过模型生成西班牙文本。seq2seq 模型由一个编码器和一个解码器组成，两者都是基于 RNN 结构。seq2seq 模型可以堆叠起来构建分布式多层模型。

6.3 LSTM 网络

当 RNN 通过非常长的数据序列进行训练时，梯度往往会变得非常大，或非常小，小到几乎接近于零。LSTM 网络通过增加用于控制对过去信息访问的门来解决梯度消失或梯度爆炸问题。LSTM 网络概念最早由 Hochreiter 和 Schmidhuber 于 1997 年提出。

阅读以下有关 LSTM 网络的研究文章，以获取有关 LSTM 网络起源的更多信息：
S. Hochreiter and J. Schmidhuber, Long Short-Term Memory, Neural Comput., vol.9, no.8, pp.1735–1780, 1997. http://www.bioinf.jku.at/ publications/older/2604.pdf.

在 RNN 中，采用重复使用学习函数 φ 的单个神经网络层，而在 LSTM 网络中，使用由四个主要函数组成的重复模块。构建 LSTM 网络的模块称为单元（cell），当长序列通过时，LSTM 单元通过选择性地学习或删除信息以便能更有效地训练模型。组成单元的函数也被称为门，因为该函数充当了传入和传出单元信息的守门人。

LSTM 模型有两种内存：
- 用 h（隐藏状态）表示的工作记忆；
- 用 c（单元状态）表示的长期记忆。

单元状态或长期记忆只通过两种线性相互作用以便从一个单元流向另一个单元。LSTM 模型利用门将信息添加到长期记忆中，或从长期记忆中删除信息。

图 6-5 为 LSTM 单元示意图。

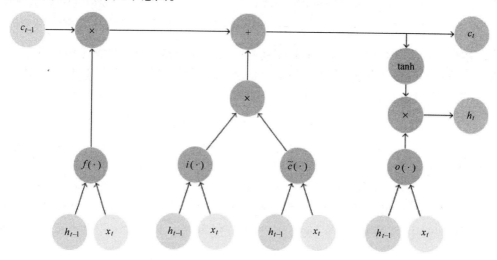

图 6-5

在 LSTM 单元中通过门的内部流动如下：

1）**遗忘门** $f()$（**或者记忆门**）：对于下式，h_{t-1} 和 x_t 作为输入流到 $f()$ 门：

$$f(\cdot) = \sigma\left(w^{(fx)} \cdot x_t + w^{(fh)} \cdot h_{t-1} + b^{(f)}\right)$$

遗忘门的功能是决定忘记哪些信息以及要记住哪些信息。这里使用 sigmoid 作为激活函数，因此，输出为 1 表示该信息被转入下一步的单元，输出为 0 表示该信息被选择性地丢弃。

2）**输入门** $i()$（**或者保存门**）：对于下式，h_{t-1} 和 x_t 作为输入流到 $i()$ 门：

$$i(\cdot) = \sigma\left(w^{(ix)} \cdot x_t + w^{(ih)} \cdot h_{t-1} + b^{(i)}\right)$$

输入门的功能是决定保存还是丢弃输入。输入功能还允许单元学习保留或丢弃候选记忆的哪一部分。

3）**候选长期记忆**：根据下式，通过 h_{t-1} 和 x_t 来计算候选长期记忆，很多时候会采用 tanh 作为激活函数：

$$\tilde{c}(\cdot) = \tanh\left(w^{(\tilde{c}x)} \cdot x_t + w^{(\tilde{c}h)} \cdot h_{t-1} + b^{(\tilde{c})}\right)$$

4）接下来，结合前面的三个计算来更新长期记忆 c_t，可表示如下：

$$c_t = c_{t-1} \times f(\cdot) + i(\cdot) \times \tilde{c}(\cdot)$$

5）**输出门** $o(\ )$（或者**焦点/注意门**）：按照下式，h_{t-1} 和 x_t 作为输入流到 $o(\ )$ 门：

$$o(\cdot) = \sigma\left(w^{(ox)} \cdot x_t + w^{(oh)} \cdot h_{t-1} + b^{(o)}\right)$$

输出门的功能是决定可以使用多少信息来更新工作记忆。

6）接下来，根据下式从长期记忆 c_t 和焦点/注意向量来更新工作记忆 h_t：

$$h_t = \varphi(c_t) \times o(\cdot)$$

其中，$\varphi(\cdot)$ 是激活函数，通常是 tanh。

6.4 GRU 网络

LSTM 网络的计算量很大，因此研究人员发现了一种几乎同样有效的 RNN 结构，称为**门控循环单元（GRU）**结构。

在 GRU 结构中，并不会使用工作记忆和长期记忆，而是只使用一种记忆，即 h（隐藏状态）。GRU 单元将信息添加到该状态记忆或通过重置和更新门从该状态的记忆中移除信息。

图 6-6 为 GRU 单元示意图（在图的下面有该图的解释）。

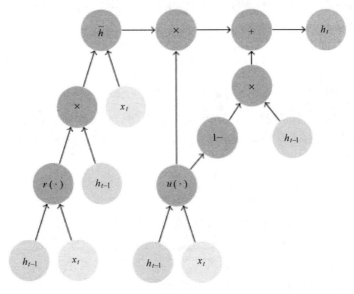

图 6-6

在 GRU 单元中，通过门的内部流程如下：

1）**更新门** $u(\ )$：输入 h_{t-1} 和 x_t 根据下式流入 $u(\ )$ 门：

$$u(\cdot) = \sigma\left(w^{(ux)} \cdot x_t + w^{(uh)} \cdot h_{t-1} + b^{(u)}\right)$$

2）重置门 r( )：输入 h_{t-1} 和 x_t 根据下式流入 r( ) 门：

$$r(\cdot) = \sigma\left(w^{(rx)} \cdot x_t + w^{(rh)} \cdot h_{t-1} + b^{(r)}\right)$$

3）候选状态记忆：根据下式，由 r( ) 门的输出、h_{t-1} 和 x_t 来计算候选长期记忆：

$$\tilde{h}(\cdot) = \tanh\left(w^{(\tilde{h}x)} \cdot x_t + w^{(\tilde{h}h)} \cdot (r_t \cdot h_{t-1}) + b^{(\tilde{h})}\right)$$

4）接下来，将前面的三个计算结合起来，得到更新的状态记忆 h_t，用下式表示：

$$h_t = \left(u_t \cdot \tilde{h}_t\right) + \left((1 - u_t) \cdot h_{t-1}\right)$$

读者可阅读以下研究论文来获得 GRU 网络的更多细节：

K. Cho, B. van Merrienboer, C. Gulcehre, D. Bahdanau, F. Bougares, H. Schwenk, and Y. Bengio, Learning Phrase Representations using RNN Encoder-Decoder for Statistical Machine Translation, 2014. https://arxiv.org/abs/1406.1078；

J. Chung, C. Gulcehre, K. Cho, and Y. Bengio, Empirical Evaluation of Gated Recurrent Neural Networks on Sequence Modeling, pp. 1–9, 2014. https://arxiv.org/abs/1412.3555。

6.5 基于 TensorFlow 的 RNN

在低级 TensorFlow 库中创建 RNN 模型的基本工作流程与 MLP 几乎相同：
- 首先创建形状 (None, # TimeSteps, # Features) 或者 (Batch Size, # TimeSteps, # Features) 的输入和输出占位符；
- 为输入占位符创建一个长度为 #TimeSteps 的列表，其中包含形状张量 (None, #Features) 或者 (Batch Size, # Features)；
- 通过 tf.rnn.rnn_cell 模块创建一个希望得到的 RNN 单元；
- 使用先前创建的单元和输入张量列表创建静态或动态 RNN；
- 创建输出权重和偏差变量，并定义损失函数和优化器函数；
- 给定循环次数，使用损失函数和优化器函数训练模型。

这个基本的工作流程将在下一章的示例代码中演示。现在来看看可用于支持上面工作流程的类。

6.5.1 TensorFlow 的 RNN 单元类

tf.nn.rnn_cell 模块包含表 6-1 所示用于在 TensorFlow 中创建不同单元的类。

表 6-1

类 名	描 述
BasicRNNCell	提供简单的 RNN 单元
BasicLSTMCell	基于 http://arxiv.org/abs/1409.2329 实现的简单 LSTM RNN 单元
LSTMCell	基于 http://deeplearning.cs.cmu.edu/pdfs/Hochreiter97_lstm.pdf 和 https://research.google.com/pubs/archive/43905.pdf 实现的简单 LSTM RNN 单元
GRUCell	基于 http://arxiv.org/abs/1406.1078 实现的 GGU RNN 单元
MultiRNNCell	实现由多个简单单元连续组成的 RNN 单元

tf.contrib.rnn 模块提供了表 6-2 所示用于创建不同单元的附加类。

表 6-2

类 名	描 述
LSTMBlockCell	基于 http://arxiv.org/abs/1409.2329 实现的 LSTM RNN 单元
LSTMBlockFusedCell	基于 http://arxiv.org/abs/1409.2329 实现的块融合 LSTM RNN 单元
GLSTMCell	基于 https://arxiv.org/abs/1703.10722 实现的 LSTM 单元组
GridLSTMCell	基于 http://arxiv.org/abs/1507.01526 实现的网格 LSTM RNN 单元
GRUBlockCell	基于 http://arxiv.org/abs/1406.1078 实现的块 GRU RNN 单元
BidirectionalGridLSTMCell	仅在频率和时间上提供双向网格 LSTM
NASCell	基于 https://arxiv.org/abs/1611.01578 实现的神经架构搜索 RNN 单元
UGRNNCell	基于 https://arxiv.org/abs/1611.09913 实现的更新门 RNN 单元

6.5.2　TensorFlow 的 RNN 模型构造类

TensorFlow 提供了从 RNN 单元对象创建 RNN 模型的类。静态 RNN 类在编译时能针对时间步长添加展开单元（unrolled cells），而动态 RNN 类在运行时能针对时间步长添加展开单元。

- tf.nn.static_rnn
- tf.nn.static_state_saving_rnn
- tf.nn.static_bidirectional_rnn
- tf.nn.dynamic_rnn
- tf.nn.bidirectional_dynamic_rnn
- tf.nn.raw_rnn
- tf.contrib.rnn.stack_bidirectional_dynamic_rnn

6.5.3　TensorFlow 的 RNN 单元封装类

TensorFlow 还提供了封装其他单元类的类：

- tf.contrib.rnn.LSTMBlockWrapper
- tf.contrib.rnn.DropoutWrapper
- tf.contrib.rnn.EmbeddingWrapper
- tf.contrib.rnn.InputProjectionWrapper
- tf.contrib.rnn.OutputProjectionWrapper
- tf.contrib.rnn.DeviceWrapper
- tf.contrib.rnn.ResidualWrapper

以下链接给出了 TensorFlow 中有关 RNN 的最新文档：https://www.tensorflow.org/api_guides/python/contrib.rnn。

6.6 基于 Keras 的 RNN

与 TensorFlow 相比，在 Keras 中创建 RNN 要容易得多。正如在第 3 章中了解到的，Keras 提供了用于创建 RNN 的函数和序列 API。要构建 RNN 模型，必须通过 kera.layers.recurrent 模块来添加网络层。在 keras.layers.recurrent 模块中，Keras 提供了以下几种循环层：

- SimpleRNN
- LSTM
- GRU

有状态模型

Keras 循环层还支持在批次之间保存状态的 RNN 模型，可以通过将有状态参数设为 True 来创建有状态的 RNN、LSTM 或 GRU 模型。对于有状态模型，为输入指定的批次大小必须是固定值。在有状态模型中，从批次训练中学习到的隐藏状态将被重新用于下一批次。如果想在训练期间的某个时候重置内存，可通过调用 model.reset_states( ) 或 layer.reset_states( ) 函数来完成。

下一章将给出使用 Keras 构建 RNN 的例子。

有关 Keras 中的递归层的最新文档可以在以下链接中找到：https://keras.io/layers/recurrent/。

6.7 RNN 的应用领域

下面是一些经常使用 RNN 的应用领域：

- **自然语言建模**：RNN 模型已用于自然语言处理（NLP），以完成自然语言理解和自然语言生成任务。在 NLP 中，给 RNN 模型一个单词序列，它能预测另一个单词序列。因此，训练过的模型可以用于生成词序列，这个领域被称为文本生成。例如，生成故事和剧本。NLP 的另一个领域是语言翻译，在给定一种语言的单词序列的情况下，该模型能预测另一种语言的单词序列。

- **声音和语音识别**：RNN 模型在建立从音频数据中学习模型有很大用处。在语音识别中，为 RNN 模型提供音频数据，它能预测语音段的序列。这可以用来训练模型以识别语音命令，甚至可与聊天机器人对话。

- **图像/视频描述或生成摘要**：RNN 模型可与 CNN 模型相结合来描述图像和视频中的元素。这些描述也可以用于生成图像和视频的摘要。

- **时间序列数据**：这是最重要的应用，RNN 对时间序列数据非常有用，大多数传感器和系统生成与时间顺序非常相关的数据。RNN 模型非常适合查找时间数据模式和预测时间数据。

以下链接有助于了解更多有关 RNN 的信息：

http://karpathy.github.io/2015/05/21/rnn-effectiveness/;

http://colah.github.io/posts/2015-08-Understanding-LSTMs/;

http://www.wildml.com/2015/09/recurrent-neural-networks-tutorial-part-1-introduction-to-rnns/;

https://r2rt.com/written-memories-understanding-deriving-and- extending-the-lstm.html。

6.8 将基于 Keras 的 RNN 用于 MNIST 数据

尽管 RNN 主要用于序列数据，但也可用于图像数据。众所周知，图像具有两个维度：高度和宽度。现在将其中一个维度看作时间步长，另一维度则视为特征。对于 MNIST 数据，图像大小为 28×28 像素，因此可以将 MNIST 图像想象成 28 个时间步长，每个时间步长具有 28 个特征。

下一章将给出基于时间序列和文本数据的例子，但这里先基于 Keras 来构建和训练一个 RNN，以便读者能快速了解构建和训练 RNN 模型的过程。

下面的代码可以在 Jupyter 笔记本 ch-06_RNN_MNIST_Keras 中找到。

导入所需的模块：

```
import keras
from keras.models import Sequential
from keras.layers import Dense, Activation
from keras.layers.recurrent import SimpleRNN
from keras.optimizers import RMSprop
from keras.optimizers import SGD
```

获取 MNIST 数据并将数据从一维结构中的 784 像素转换为二维结构中的 28×28 像素：

```
from tensorflow.examples.tutorials.mnist import input_data
mnist = input_data.read_data_sets(os.path.join(datasetslib.datasets_root,
                                                'mnist'),
                                   one_hot=True)
X_train = mnist.train.images
X_test = mnist.test.images
Y_train = mnist.train.labels
Y_test = mnist.test.labels
n_classes = 10
n_classes = 10
X_train = X_train.reshape(-1,28,28)
X_test = X_test.reshape(-1,28,28)
```

在 Keras 中构建 SimpleRNN 模型：

```
# 创建和拟合 SimpleRNN 模型
model = Sequential()
model.add(SimpleRNN(units=16, activation='relu', input_shape=(28,28)))
model.add(Dense(n_classes))
model.add(Activation('softmax'))

model.compile(loss='categorical_crossentropy',
              optimizer=RMSprop(lr=0.01),
              metrics=['accuracy'])
model.summary()
```

该模型的信息如下：

```
Layer (type)                 Output Shape              Param #
=================================================================
simple_rnn_1 (SimpleRNN)     (None, 16)                720

dense_1 (Dense)              (None, 10)                170

activation_1 (Activation)    (None, 10)                0
=================================================================
Total params: 890
Trainable params: 890
Non-trainable params: 0
```

训练模型并输出测试数据集的分类精度：

```
model.fit(X_train, Y_train,
          batch_size=100, epochs=20)

score = model.evaluate(X_test, Y_test)
print('\nTest loss:', score[0])
print('Test accuracy:', score[1])
```

可以得到以下结果：

```
Test loss: 0.520945608187
Test accuracy: 0.8379
```

6.9 总结

通过本章可以了解循环神经网络（RNN），以及 RNN 的各种改进版本，本章详细介绍了其中两种改进版本：长短期记忆（LSTM）网络和门控循环单元（GRU）网络。本章还介绍了可用于在 TensorFlow 和 Keras 中构建 RNN 单元、模型和网络层的类，并建立了一个简单的 RNN，该网络能用于对 MNIST 数据集中的数字进行分类。

下一章将介绍如何基于时间序列数据来构建和训练 RNN 模型。

第 7 章
基于 TensorFlow 和 Keras 的 RNN 在时间序列数据中的应用

时间序列数据是一系列在不同时间间隔记录或测量的值。对于时间序列数据而言，RNN 结构是从这些数据训练模型的最佳方式。本章将使用时间序列数据集来展示如何使用 TensorFlow 和 Keras 构建 RNN 模型。

本章将介绍以下主题：
- 航空公司乘客数据集：
  - 加载 airpass 数据集；
  - 可视化 airpass 数据集。
- 使用 TensorFlow 为 RNN 模型预处理数据集。
- 使用 TensorFlow 的 RNN 模型处理时间序列数据：
  - TensorFlow 中的简单 RNN；
  - TensorFlow 中的 LSTM 网络；
  - TensorFlow 中的 GRU 网络。
- 使用 Keras 的 RNN 模型处理时间序列数据：
  - 基于 Keras 的简单 RNN；
  - 基于 Keras 的 LSTM 网络；
  - 基于 Keras 的 GRU 网络。

下面先介绍示例数据集。

 读者可按照 Jupyter 笔记本 ch-07a_RNN_TimeSeries_TensorFlow 中的相关代码来进行操作。

7.1 航空公司乘客数据集

为简洁起见，选择了一个国际航空乘客（International Airline Passenger，airpass）数据集，该数据集非常小。这些数据包含 1949 年 1 月至 1960 年 12 月每月的乘客总数。数据集中的数字是以千为单位。该数据集于 1976 年最先在 Box 和 Jenkins 的工作中使用，其中一部分数据被澳大利亚 Monash 大学的 Rob Hyndman 教授收集到了**时间序列数据库**

（TimeSeries Dataset Library，TSDL），该数据库还收集了其他类型的时间序列数据。后来，TSDL 移交给了 DataMarket（http://datamarket.com）。

 可以从下面的链接下载数据集：https://datamarket.com/data/set/22u3/international-airline-passengers-monthly-totals-in-thousands-jan-49-dec-60。

7.1.1 加载 airpass 数据集

将数据集作为 CSV 文件保存在数据集根目录（~/ datasets）的 ts-data 文件夹中，并使用以下命令将数据加载到 pandas 数据框中：

```
filepath = os.path.join(datasetslib.datasets_root,
                        'ts-data',
                        'international-airline-passengers-cleaned.csv'
                        )
dataframe = pd.read_csv(filepath,usecols=[1],header=0)
dataset = dataframe.values
dataset = dataset.astype(np.float32)
```

从 NumPy 数组中提取数据框中的值并将其转换为 np.float32：

```
dataset = dataframe.values
dataset = dataset.astype(np.float32)
```

7.1.2 可视化 airpass 数据集

下面来看看数据集的形状：

```
plt.plot(dataset,label='Original Data')
plt.legend()
plt.show()
```

绘制 airpass 数据集的结果如图 7-1 所示。

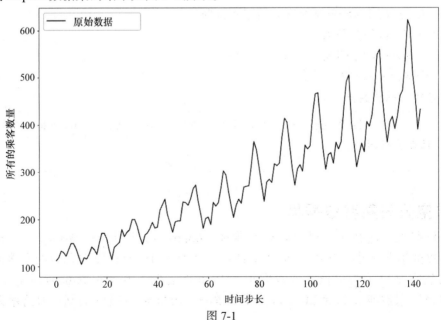

图 7-1

7.2 使用 TensorFlow 为 RNN 模型预处理数据集

在学习模型之前，可通过 MinMax 缩放函数来归一化数据集，使数据的值都介于 0 和 1 之间。读者可以尝试对数据采用不同的缩放方法，具体取决于数据的性质。

```
# 归一化数据集
scaler = skpp.MinMaxScaler(feature_range=(0, 1))
normalized_dataset = scaler.fit_transform(dataset)
```

使用自定义函数将数据集分解为训练数据集和测试数据集。由于混排（shuffle）会打破数据的时间顺序，因此，只对数据集进行拆分。维护数据的顺序对训练时间序列模型很重要。

```
train,test=tsu.train_test_split(normalized_dataset,train_size=0.67)
```

接下来将把训练数据集和测试数据集转换为监督机器学习的数据集。读者可试着去理解监督学习集的含义，假设有一个序列数据:1,2,3,4,5，想要了解生成数据集的概率分布。为了做到这一点，可以假设某个时间点 t 的值是基于时间点 $t-1$ 到 $t-k$ 的值得到的，其中 k 是窗口大小。为了简单起见，假设窗口大小为 1，因此，时间点 t 处的值（称为输入特征）是基于 $t-1$ 处的值（称为目标）得到的。对所有的时间点都是这样，于是就能得到表 7-1 的内容。

表 7-1

输入数据或特征	输出值或目标
1	2
2	3
3	4
4	5

在上面的例子中，特征和目标都只有一个变量。目标值为一个变量时，称为单变量时间序列。可用同样的方式定义多变量时间序列。用 x 代表输入特征，用 y 代表输出目标。

为了将航空乘客数据转换为监督机器学习的数据，设置以下超参数：

1）设置用于学习或预测下一时间点的过去时间点的数量：

```
n_x=1
```

2）设置要学习或预测的将来时间点的数量：

```
n_y=1
```

3）设置用于学习的变量数量；因为当前的例子是单变量，所以设置为 1：

```
n_x_vars = 1
```

4）设置要预测的变量数量；因为当前的例子是单变量，所以设置为 1：

```
n_y_vars = 1
```

5）最后，通过应用本节开头介绍的逻辑将训练和测试数据集分配给变量 X 和 Y：

```
X_train, Y_train, X_test, Y_test = tsu.mvts_to_xy(train,
                                    test,n_x=n_x,n_y=n_y)
```

现在数据已经经过预处理并可以输入到模型中，下面使用 TensorFlow 来构建简单 RNN 模型。

7.3 TensorFlow 中的简单 RNN

在 TensorFlow 中定义和训练简单 RNN 的工作流程如下：

1）定义模型的超参数：

```
state_size = 4
n_epochs = 100
n_timesteps = n_x
learning_rate = 0.1
```

这里的新超参数是 state_size，它表示 RNN 单元权重向量的数量。

2）为模型的 X 和 Y 定义占位符。X 占位符的形状为（batch_size, number_of_input_timesteps, number_of_inputs），Y 占位符的形状为（batch_size, number_of_output_timesteps, number_of_outputs）。对于 batch_size，使用 None，以便稍后可以输入任何大小的批次。

```
X_p = tf.placeholder(tf.float32, [None, n_timesteps, n_x_vars],
    name='X_p')
Y_p = tf.placeholder(tf.float32, [None, n_timesteps, n_y_vars],
    name='Y_p')
```

3）将输入占位符 X_p 转换为长度等于时间点数量的张量列表，在本例中为 n_x 或 1：

```
# 制作长度为n_timesteps的张量列表
rnn_inputs = tf.unstack(X_p,axis=1)
```

4）使用 tf.nn.rnn_cell.BasicRNNCell 创建一个简单的 RNN 单元：

```
cell = tf.nn.rnn_cell.BasicRNNCell(state_size)
```

5）TensorFlow 提供 static_rnn 和 dynamic_rnn 方法（以及其他方法）来分别创建静态 RNN 和动态 RNN。下面创建一个静态 RNN：

```
rnn_outputs, final_state = tf.nn.static_rnn(cell,
                                rnn_inputs,
                                dtype=tf.float32
                                )
```

静态 RNN 在编译时创建单元，即展开循环。动态 RNN 在运行时创建单元，即展开循环。在本章中，只展示创建静态 RNN 的例子，但是，一旦有了静态 RNN 的知识，就应该可以创建动态 RNN。

static_rnn 方法需要的参数有：

- cell：之前定义的基本 RNN 单元对象，可以是另一种类型的单元，本章会进一步讨论。
- rnn_inputs：形状张量列表（batch_size, number_of_inputs）。
- dtype：初始状态和希望输出的数据类型。

6）为预测层定义权重和偏差参数：

```
W = tf.get_variable('W', [state_size, n_y_vars])
b = tf.get_variable('b', [n_y_vars],
    initializer=tf.constant_initializer(0.0))
```

7）将预测层定义为一个密集的线性层：

```
predictions = [tf.matmul(rnn_output, W) + b \
               for rnn_output in rnn_outputs]
```

8）以张量形式输出 Y，并将其转换为张量列表：

```
y_as_list = tf.unstack(Y_p, num=n_timesteps, axis=1)
```

9）将损失函数定义为预测标签与实际标签之间的均方误差：

```
mse = tf.losses.mean_squared_error
losses = [mse(labels=label, predictions=prediction)
          for prediction, label in zip(predictions, y_as_list)
         ]
```

10）将总损失函数定义为所有预测的时间点的平均损失：

```
total_loss = tf.reduce_mean(losses)
```

11）定义优化器来最小化 total_loss：

```
optimizer =
tf.train.AdagradOptimizer(learning_rate).minimize(total_loss)
```

12）现在定义了模型、损失函数和优化器函数，接下来，训练模型并计算训练损失值：

```
with tf.Session() as tfs:
    tfs.run(tf.global_variables_initializer())
    epoch_loss = 0.0
    for epoch in range(n_epochs):
        feed_dict={X_p: X_train.reshape(-1, n_timesteps,
                                        n_x_vars),
                   Y_p: Y_train.reshape(-1, n_timesteps,
                                        n_x_vars)
                  }
        epoch_loss,y_train_pred,_=tfs.run([total_loss,predictions,
                            optimizer], feed_dict=feed_dict)
    print("train mse = {}".format(epoch_loss))
```

这会得到以下结果：

```
train mse = 0.0019413739209994674
```

13）用测试数据来测试模型，会得到如下结果：

```
feed_dict={X_p: X_test.reshape(-1, n_timesteps,n_x_vars),
           Y_p: Y_test.reshape(-1, n_timesteps,n_y_vars)
          }
test_loss, y_test_pred = tfs.run([total_loss,predictions],
                        feed_dict=feed_dict
                        )
```

```
print('test mse = {}'.format(test_loss))
print('test rmse = {}'.format(math.sqrt(test_loss)))
```

在测试数据上得到以下均方误差和均方根误差：

```
test mse = 0.008790395222604275
test rmse = 0.09375710758446143
```

这个结果相当不错。

 这是一个非常简单的例子，只用一个变量值就可以预测一个时间点上的结果。在现实生活中，输出会受多个特征影响，并且需要预测多个时间点。后一类问题称为多变量多时间预测问题，这是一个非常活跃的研究领域，可使用 RNN 获得更好的预测结果。

现在，重新调整预测值和原始值，并对原始值绘图（请在 Jupyter 笔记本中查找代码）。会得到图 7-2 所示结果。

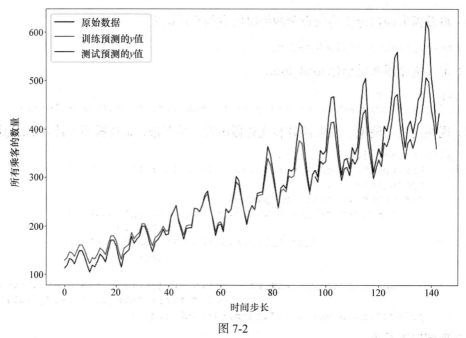

图 7-2

在这个简单的例子中，预测的数据几乎与原始数据完全一样，这是非常好的预测结果。对这种精确预测的一种解释是单个时间点的预测是基于来自上一个时间点的单变量预测，因此总会在上一个值的附近。

这个例子只是为了展示如何在 TensorFlow 中创建一个 RNN。现在使用 RNN 的改进方法重新创建相同的示例。

7.4　TensorFlow 中的 LSTM 网络

简单 RNN 会出现梯度爆炸和梯度消失问题，当出现这些问题时，RNN 就不能工作，因此需要使用改进 RNN，例如 LSTM 网络。TensorFlow 提供了创建 LSTM RNN 的 API。

对于上一节所展示的例子，若要将简单 RNN 更改为 LSTM 网络，只需要更改单元类型即可：

```
cell = tf.nn.rnn_cell.LSTMCell(state_size)
```

在 LSTM 单元中创建 RNN 的其余代码与在 TensorFlow 中完全相同。

在 Jupyter 笔记本 ch-07a_RNN_TimeSeries_TensorFlow 中提供了 LSTM 模型的完整代码。

但是，对于 LSTM 网络，必须运行 600 次循环才能使结果更接近基本的 RNN。原因是 LSTM 网络有更多的参数需要学习，因此需要更多的迭代次数。就这个简单的例子而言，这样的需求似乎高了一些，但对于较大的数据集，LSTM 网络会得到比 RNN 更好的结果。

基于 LSTM 架构的模型所输出的结果（见图 7-3）如下：

```
train mse = 0.0020806745160371065
test mse = 0.01499235536903143
test rmse = 0.12244327408653947
```

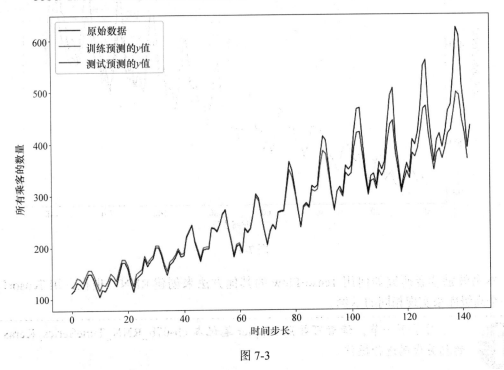

图 7-3

7.5 TensorFlow 中的 GRU 网络

要将上一节中的 LSTM 网络示例更改为 GRU 网络，只需按如下所示更改单元类型即可，剩下的事情全部由 TensorFlow 来处理：

```
cell = tf.nn.rnn_cell.GRUCell(state_size)
```

在 Jupyter 笔记本 ch-07a_RNN_TimeSeries_TensorFlow 中提供了 GRU 模型的完整代码。

对于 airpass 数据集，GRU 模型在同样数量的时间点下会表现出更好的性能。实际上，GRU 模型和 LSTM 模型都表现出不错的性能。就执行速度而言，GRU 模型的训练速度和预测速度会更快。

Jupyter 笔记本提供了 GRU 模型的完整代码。基于 GRU 模型会得到如下结果（见图 7-4）：

```
train mse = 0.0019633215852081776
test mse = 0.014307591132819653
test rmse = 0.11961434334066987
```

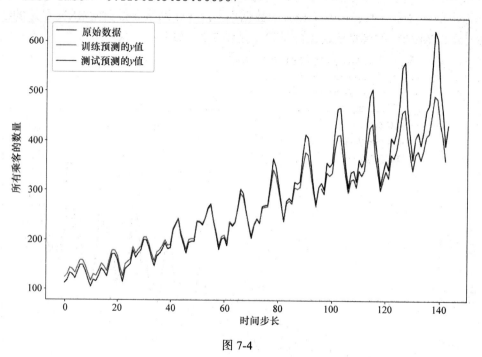

图 7-4

本书鼓励读者研究如何用 TensorFlow 的其他方法来创建 RNN。现在，用 TensorFlow 的一个高级库来实现相同的示例。

 对于下一节，读者可按照 Jupyter 笔记本 ch-07b_RNN_TimeSeries_Keras 中的相关代码进行操作。

7.6 使用 Keras 为 RNN 模型预处理数据集

与使用低级 TensorFlow 类和方法相比，在 Keras 中构建 RNN 要简单得多。在使用 Keras 之前，需要预处理数据来获得监督时间序列数据集：X_train、Y_train、X_test、Y_test。

这里的预处理与前面有所不同。对于 Keras 来说，输入形状必须为 (samples, time steps, features)。由于数据转换为监督机器学习格式，当重新变换数据时，可以将时间步长设置为 1，此时，所有输入时间点都是特征，或者可以将时间点设置为实际的时间步数，从而在每个时间点上提供特征集。换句话说，可通过下面的方法对之前获得的 X_train 和 X_test 进行重新定义。

方法 1：具有 1 个特征的 *n* 个时间点：

```
X_train.reshape(X_train.shape[0], X_train.shape[1],1)
```

方法 2：具有 *n* 个特征的 1 个时间点：

```
X_train.reshape(X_train.shape[0], 1, X_train.shape[1])
```

本章将使用特征大小为 1 的数据集，因为只有一个变量作为输入：

```
# 重塑输入为 [samples, time steps, features]
X_train = X_train.reshape(X_train.shape[0], X_train.shape[1],1)
X_test = X_test.reshape(X_test.shape[0], X_train.shape[1], 1)
```

7.7 基于 Keras 的简单 RNN

通过将 SimpleRNN 层与内部神经元的数量以及输入张量的形状（但不包括样本数量维数）相加，可以在 Keras 中方便地构建 RNN 模型。以下代码将创建、编译并训练 SimpleRNN：

```
# 创建和训练 SimpleRNN 模型
model = Sequential()
model.add(SimpleRNN(units=4, input_shape=(X_train.shape[1],
    X_train.shape[2])))
model.add(Dense(1))
model.compile(loss='mean_squared_error', optimizer='adam')
model.fit(X_train, Y_train, epochs=20, batch_size=1)
```

由于数据集很小，因此设置 batch_size 为 1，并训练 20 次，但对于较大的数据集，需要调整相关的参数值。

该模型的结构如下：

```
Layer (type)                    Output Shape          Param #
=================================================================
simple_rnn_1 (SimpleRNN)        (None, 4)             24
_____
dense_1 (Dense)                 (None, 1)             5
=================================================================
Total params: 29
Trainable params: 29
Non-trainable params: 0
```

训练结果如下：

```
Epoch 1/20
95/95 [==============================] - 0s - loss: 0.0161
Epoch 2/20
95/95 [==============================] - 0s - loss: 0.0074
Epoch 3/20
95/95 [==============================] - 0s - loss: 0.0063
Epoch 4/20
95/95 [==============================] - 0s - loss: 0.0051

-- epoch 5 to 14 removed for the sake of brevity --

Epoch 14/20
95/95 [==============================] - 0s - loss: 0.0021
Epoch 15/20
95/95 [==============================] - 0s - loss: 0.0020
Epoch 16/20
95/95 [==============================] - 0s - loss: 0.0020
Epoch 17/20
95/95 [==============================] - 0s - loss: 0.0020
Epoch 18/20
95/95 [==============================] - 0s - loss: 0.0020
Epoch 19/20
95/95 [==============================] - 0s - loss: 0.0020
Epoch 20/20
95/95 [==============================] - 0s - loss: 0.0020
```

损失函数值为 0.0161，然后降低为 0.0020。接下来，会进行预测并重新调整预测结果和原来的值。使用 Keras 提供的函数来计算均方根误差：

```python
from keras.losses import mean_squared_error as k_mse
from keras.backend import sqrt as k_sqrt
import keras.backend as K

# 进行预测
y_train_pred = model.predict(X_train)
y_test_pred = model.predict(X_test)

# 调整预测值
y_train_pred = scaler.inverse_transform(y_train_pred)
y_test_pred = scaler.inverse_transform(y_test_pred)

# 调整原始值
y_train_orig = scaler.inverse_transform(Y_train)
y_test_orig = scaler.inverse_transform(Y_test)

# 计算均方根误差
trainScore = k_sqrt(k_mse(y_train_orig[:,0],
                          y_train_pred[:,0])
                    ).eval(session=K.get_session())
print('Train Score: {0:.2f} RMSE'.format(trainScore))

testScore = k_sqrt(k_mse(y_test_orig[:,0],
                         y_test_pred[:,0])
                   ).eval(session=K.get_session())
print('Test Score: {0:.2f} RMSE'.format(testScore))
```

得到以下结果（见图 7-5）：

```
Train Score: 23.27 RMSE
Test Score: 54.13 RMSE
```

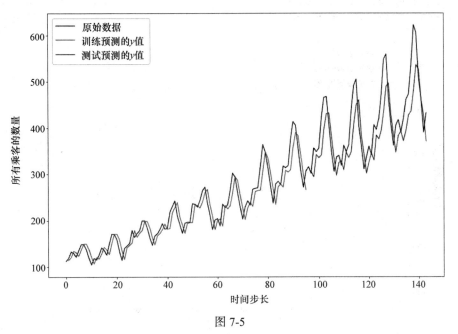

图 7-5

正如所看到的，这并不像直接利用 TensorFlow 得到的那样完美，这种差异与超参数的值有关。读者可尝试调整超参数的值来得到比 Keras 模型更好的结果。

7.8 基于 Keras 的 LSTM 网络

创建 LSTM 模型只需要添加 LSTM 层即可，如下所示：

`model.add(LSTM(units=4, input_shape=(X_train.shape[1], X_train.shape[2])))`

模型结构如下所示：

```
Layer (type)                 Output Shape              Param #
=================================================================
lstm_1 (LSTM)                (None, 4)                 96
_____
dense_1 (Dense)              (None, 1)                 5
=================================================================
Total params: 101
Trainable params: 101
Non-trainable params: 0
```

LSTM 模型的完整代码在 Jupyter 笔记本 ch-07b_RNN_TimeSeries_Keras 中可找到。
由于 LSTM 模型有更多需要训练的参数，对于相同的迭代次数（20 次迭代），将得到更高的误差值。读者可以研究如何通过调整步长数和其他超参数值来得到更好的结果（见图 7-6）。

```
Train Score: 32.21 RMSE
Test Score: 84.68 RMSE
```

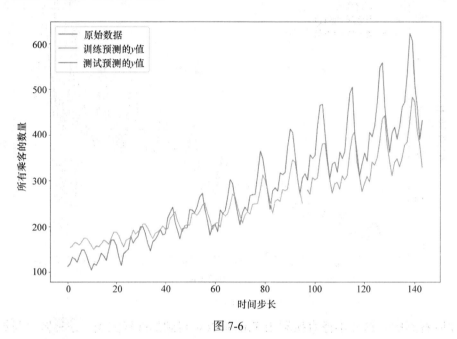

图 7-6

7.9 基于 Keras 的 GRU 网络

使用 Keras 的优势在于可以轻松创建模型。就像 LSTM 模型一样，创建 GRU 模型只需要添加 GRU 层即可，具体实现如下：

```
model.add(GRU(units=4, input_shape=(X_train.shape[1], X_train.shape[2])))
```

模型结构如下：

```
Layer (type)                 Output Shape              Param #
=================================================================
gru_1 (GRU)                  (None, 4)                 72
_____
dense_1 (Dense)              (None, 1)                 5
=================================================================
Total params: 77
Trainable params: 77
Non-trainable params: 0
```

GRU 模型的完整代码可在 Jupyter 笔记本 ch-07b_RNN_TimeSeries_Keras 中找到。

正如预料的那样，GRU 模型与 LSTM 模型在性能上几乎一样，读者可以尝试使用不同的超参数值来优化此模型（见图 7-7）：

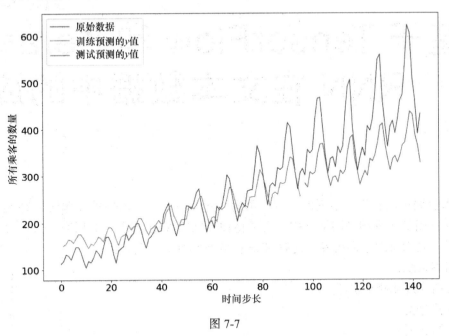

图 7-7

7.10 总结

时间序列数据是基于序列的数据,因此 RNN 模型是从时间序列数据中学习到的结构。本章介绍了如何使用 TensorFlow(低级库)和 Keras(高级库)创建不同类型的 RNN 模型。这里只创建了 SimpleRNN、LSTM 和 GRU 模型,但读者也可以使用 TensorFlow 和 Keras 来创建其他 RNN 模型。

下一章在本章和前面的基础上,为各种自然语言处理(NLP)任务创建基于文本数据的 RNN 模型。

第8章 基于 TensorFlow 和 Keras 的 RNN 在文本数据中的应用

文本数据可以看作一系列的字符、单词、句子或段落。循环神经网络（RNN）已被证明是非常有用的序列神经网络结构。为了将神经网络模型应用于自然语言处理（NLP），可将文本看作是由有序单词构成，这对 NLP 任务非常有效，例如：
- 问答系统；
- 会话代理或聊天机器人；
- 文档分类；
- 情绪分析；
- 图像摘要或生成文本描述；
- 命名实体识别；
- 语音识别和标注。

基于深度学习的 NLP 是一个广阔的领域，很难通过一章将全部内容讲清楚。因此，本章将基于 Tensorflow 和 Keras 来介绍该领域最普遍和最重要的例子。一旦掌握了本章的内容，就可以去研究 NLP 的其他领域。

本章将介绍以下主题：
- 词向量表示；
- 为 word2vec 模型准备数据；
- 使用 TensorFlow 和 Keras 的 skip-gram 模型；
- 使用 t-SNE 可视化单词嵌入；
- 使用 TensorFlow 和 Keras 中的 LSTM 模型生成文本。

8.1 词向量表示

为了从文本数据中学习神经网络模型的参数，首先必须将文本或自然语言数据转换为神经网络可使用的格式。输入给神经网络的文本通常具有数字向量的形式。将原始文本数据转换为数字向量的算法被称为词向量嵌入算法。

一种流行的词向量嵌入方法是在 MNIST 图像分类中所采用过的**独热编码**（one-hot encoding）。假设文本数据集由 60 000 个字典单词组成。则每个单词可以用一个独热编码向量表示，该向量的长度为 60 000，除了表示该单词的元素为 1 以外，其他所有元素的值都为 0。

但独热编码方法有缺点。首先，对于具有大量单词的文本，基于独热编码的词向量的维度会非常高。其次，人们无法通过独热编码向量来得到单词的相似性。例如，假设 cat 和 kitten 的向量分别为 [1 0 0 0 0 0] 和 [0 0 0 0 0 1]，但两个向量并不相似。

还有其他基于语料库的方法可用来将文本转换为数字向量，例如：
- 词频 - 逆文本频率（TF-IDF）；
- 潜在语义分析（LSA）；
- 主题建模。

最近，用数字向量表示单词的重点已转移到基于分布假说的方法上，这意味着具有相似语义的单词会出现在相似的背景下。

word2vec 和 GloVe 是两种广泛使用的词向量方法，本章会使用 word2vec。正如前面所介绍的那样，独热编码得到的词向量有很高的维数，而使用 word2vec 得到的词向量维度要低得多。

可用两种结构来构建 word2vec 模型：
- **连续词袋**（Continuous Bag of Words，CBOW）：通过学习给定上下文单词的中心单词的概率分布来训练模型。因此，给定一组具有上下文的词语，该模型会以在高中语文课上的填空方式来预测中心词。CBOW 结构适用于拥有较小词汇表的数据集。
- **Skip-gram**：训练模型以学习给定中心词的上下文词的概率分布。因此，给定一个中心词，该模型会以在高中语文课上所做的完成句子的方式预测语境词。

例如，考虑一下这句话：

Vets2data.org is a non-profit for educating the US Military Veterans Community on Artificial Intelligence and Data Science.

在 CBOW 结构中，对于给定单词 Military 和 Community，该模型学习单词 Veterans 的概率；在 skip-gram 结构中，对于给定单词 Veterans，该模型学习单词 Military 和 Community 的概率。

word2vec 模型以无监督方式从文本语料库中学习词向量。文本语料库被分成上下文词对和目标词。虽然这些单词对是真正的对，但同时将不配对的上下文词随机配对，生成假的单词对，因此会在数据中产生噪声。训练分类器的参数用于从假的单词对中区分出真的单词对。该分类器的参数成为 word2vec 模型或词向量。

关于 word2vec 背后的数学理论，可参考以下论文：

Mikolov, T., I. Sutskever, K. Chen, G. Corrado, and J. Dean. Distributed Representations of Words and Phrases and Their Compositionality. Advances in Neural Information Processing Systems, 2013, pp. 3111–3119.

Mikolov, T., K. Chen, G. Corrado, and J. Dean. Efficient Estimation of Word Representations in Vector Space. arXiv, 2013, pp. 1–12.

Rong, X. word2vec Parameter Learning Explained. arXiv:1411.2738, 2014, pp. 1–19.

Baroni, M., G. Dinu, and G. Kruszewski. Don't Count, Predict! A Systematic Comparison of Context-Counting vs. Context-Predicting Semantic Vectors. 2014.

读者可沿着 word2vec 的思路来学习 GloVe，争取能将其应用于文本数据中。

有关 GLoVe 算法的更多信息可以从以下论文中了解到：
Pennington, J., R. Socher, and C. Manning. GloVe: Global Vectors for Word Representation. 2014.

下面通过在 TensorFlow 和 Keras 中创建词向量来理解 word2vec 模型。

Jupyter 笔记本 ch-08a_Embeddings_in_TensorFlow_and_Keras 中包含了接下来要学习的几部分代码。

8.2 为 word2vec 模型准备数据

本节将使用流行的 PTB 和 text8 数据集进行演示。

Penn Treebank（PTB）数据集是 UPenn 进行的 Penn Treebank 项目的副产品（https://catalog.ldc.upenn.edu/ldc99t42）。PTB 项目团队从 3 年的《华尔街日报》报道中提取了约 100 万字，并以 Treebank II 风格对其进行了注释。PTB 数据集有两种版本：基本示例（大小约为 35MB）和高级示例（大小约为 235MB）。使用 929K 单词组成的简单数据集进行训练，73K 单词数据用于验证，82K 单词用于测试。当然，鼓励读者探索高级数据集。有关 PTB 数据集的更多详细信息，请访问以下链接：http://www.fit.vutbr.cz/~imikolov/rnnlm/simple-examples.tgz。

PTB 数据集可以从以下链接下载：
http://www.fit.vutbr.cz/~imikolov/rnnlm/rnn-rt07-example.tar.gz。

text8 数据集是大小约为 1 GB 的维基百科数据，经过转储之后得到较短且经过清理的版本。以下链接解释了 text8 数据集的创建过程：http://mattmahoney.net/dc/textdata.html。

text8 数据集可以从以下链接下载：http://mattmahoney.net/dc/text8.zip。

数据集可使用 datasetslib 中的 load_data 方法加载。load_data() 函数会执行以下操作：

1）如果本地不可用，会从数据集的 URL 下载数据集。

2）由于 PTB 数据有三个文件，首先从训练文件中读取文本，而对于 text8 则从归档中读取第一个文件。

3）将训练文件中的单词转换成一个词汇表，为每个单词分配唯一编号（word-id），将其存储在 word2id 集合中，并准备反向字典，因此可以从 ID 中查找单词，将其存储在 id2word 集合中。

4）使用 word2id 集合将文本文件转换为 ID 序列。

5）因此，在 load_data 结束时，训练数据集中会有数字序列，并且在 id2word 集合中有一个 ID 到单词的映射。

下面介绍如何加载 text8 和 PTB 数据集。

8.2.1 加载和准备 PTB 数据集

首先导入模块并加载数据，具体操作如下：

```
from datasetslib.ptb import PTBSimple
ptb = PTBSimple()
# 加载数据，将单词转换为ids,将文件转换为ids列表
ptb.load_data()
print('Train :',ptb.part['train'][0:5])
print('Test: ',ptb.part['test'][0:5])
print('Valid: ',ptb.part['valid'][0:5])
print('Vocabulary Length = ',ptb.vocab_len)
```

每个数据集的前 5 个元素以及词汇长度如下：

```
Train : [9970, 9971, 9972, 9974, 9975]
Test:   [102, 14, 24, 32, 752]
Valid:  [1132, 93, 358, 5, 329]
Vocabulary Length =  10000
```

将上下文窗口设置为两个单词并获取 CBOW 对：

```
ptb.skip_window=2
ptb.reset_index_in_epoch()
# 在CBOW中输入是上下文单词，输出是目标单词
y_batch, x_batch = ptb.next_batch_cbow()

print('The CBOW pairs : context,target')
for i in range(5 * ptb.skip_window):
    print('(', [ptb.id2word[x_i] for x_i in x_batch[i]],
          ',', y_batch[i], ptb.id2word[y_batch[i]], ')')
```

输出的结果为：

```
The CBOW pairs : context,target
( ['aer', 'banknote', 'calloway', 'centrust'] , 9972 berlitz )
( ['banknote', 'berlitz', 'centrust', 'cluett'] , 9974 calloway )
( ['berlitz', 'calloway', 'cluett', 'fromstein'] , 9975 centrust )
( ['calloway', 'centrust', 'fromstein', 'gitano'] , 9976 cluett )
( ['centrust', 'cluett', 'gitano', 'guterman'] , 9980 fromstein )
( ['cluett', 'fromstein', 'guterman', 'hydro-quebec'] , 9981 gitano )
( ['fromstein', 'gitano', 'hydro-quebec', 'ipo'] , 9982 guterman )
( ['gitano', 'guterman', 'ipo', 'kia'] , 9983 hydro-quebec )
( ['guterman', 'hydro-quebec', 'kia', 'memotec'] , 9984 ipo )
( ['hydro-quebec', 'ipo', 'memotec', 'mlx'] , 9986 kia )
```

查看 skip-gram 对：

```
ptb.skip_window=2
ptb.reset_index_in_epoch()
# 在skip-gram 中输入是目标单词，输出是上下文单词
```

```
            x_batch, y_batch = ptb.next_batch()

            print('The skip-gram pairs : target,context')
            for i in range(5 * ptb.skip_window):
                print('(',x_batch[i], ptb.id2word[x_batch[i]],
                    ',', y_batch[i], ptb.id2word[y_batch[i]],')')
```

输出结果为：

```
The skip-gram pairs : target,context
( 9972 berlitz , 9970 aer )
( 9972 berlitz , 9971 banknote )
( 9972 berlitz , 9974 calloway )
( 9972 berlitz , 9975 centrust )
( 9974 calloway , 9971 banknote )
( 9974 calloway , 9972 berlitz )
( 9974 calloway , 9975 centrust )
( 9974 calloway , 9976 cluett )
( 9975 centrust , 9972 berlitz )
( 9975 centrust , 9974 calloway )
```

8.2.2　加载和准备 text8 数据集

现在对 text8 数据集执行相同的加载和预处理步骤：

```
from datasetslib.text8 import Text8
text8 = Text8()
text8.load_data()
# 加载数据，将单词转换为ids，将文件转换为ids列表
print('Train:', text8.part['train'][0:5])
print('Vocabulary Length = ',text8.vocab_len)
```

可以看到大约是 254 000 个单词：

```
Train: [5233, 3083, 11, 5, 194]
Vocabulary Length =  253854
```

　　一些教程通过查找最常用的单词或将词汇量大小截断为 10 000 个单词来使用此数据集。但是，本章使用了 text8 数据集的第一个文件中的所有数据集和词汇。

准备 CBOW 对：

```
text8.skip_window=2
text8.reset_index_in_epoch()
# 在CBOW中输入是上下文单词，输出是目标单词
y_batch, x_batch = text8.next_batch_cbow()

print('The CBOW pairs : context,target')
for i in range(5 * text8.skip_window):
    print('(', [text8.id2word[x_i] for x_i in x_batch[i]],
        ',', y_batch[i], text8.id2word[y_batch[i]], ')')
```

输出的结果为：

```
The CBOW pairs : context,target
( ['anarchism', 'originated', 'a', 'term'] , 11 as )
( ['originated', 'as', 'term', 'of'] , 5 a )
( ['as', 'a', 'of', 'abuse'] , 194 term )
( ['a', 'term', 'abuse', 'first'] , 1 of )
( ['term', 'of', 'first', 'used'] , 3133 abuse )
( ['of', 'abuse', 'used', 'against'] , 45 first )
( ['abuse', 'first', 'against', 'early'] , 58 used )
( ['first', 'used', 'early', 'working'] , 155 against )
( ['used', 'against', 'working', 'class'] , 127 early )
( ['against', 'early', 'class', 'radicals'] , 741 working )
```

准备 skip-gram 对：

```
text8.skip_window=2
text8.reset_index_in_epoch()
# 在skip-gram中输入是目标单词，输出是上下文单词
x_batch, y_batch = text8.next_batch()

print('The skip-gram pairs : target,context')
for i in range(5 * text8.skip_window):
    print('(',x_batch[i], text8.id2word[x_batch[i]],
          ',', y_batch[i], text8.id2word[y_batch[i]],')')
```

输出的结果为：

```
The skip-gram pairs : target,context
( 11 as , 5233 anarchism )
( 11 as , 3083 originated )
( 11 as , 5 a )
( 11 as , 194 term )
( 5 a , 3083 originated )
( 5 a , 11 as )
( 5 a , 194 term )
( 5 a , 1 of )
( 194 term , 11 as )
( 194 term , 5 a )
```

8.2.3 准备小的验证集

为了演示该示例，创建一个由 8 个单词组成的小验证集，其中，每个单词是从 word-id 在 0 到 10 × 8 之间随机选取的。

```
valid_size = 8
x_valid = np.random.choice(valid_size * 10, valid_size, replace=False)
print(x_valid)
```

将以下内容作为验证集：

```
valid:   [64 58 59 4 69 53 31 77]
```

通过输出验证集中 5 个最接近的单词来演示词嵌入的结果。

8.3 使用 TensorFlow 的 skip-gram 模型

现在已经准备好了训练数据和验证数据，需要用 TensorFlow 创建 skip-gram 模型。

首先定义超参数：

```
batch_size = 128
embedding_size = 128
skip_window = 2
n_negative_samples = 64
ptb.skip_window=2
learning_rate = 1.0
```

- batch_size 表示在单个批处理中输入算法的目标和上下文单词对的数量；
- embedding_size 表示每个单词对应的向量维度；
- ptb.skip_window 表示在两个方向上目标词要考虑的上下文中的单词数；
- n_negative_samples 表示由 NCE 损失函数生成的负样本数，本章会对此作进一步解释。

> 在一些教程（包括 TensorFlow 文档中的一个教程）中，使用了另外一个参数 num_skips。在这样的教程中，作者选择 (target, context) 对的数量为 num_skips。例如，如果 skip_window 为 2，则 (target, context) 对的总数将为 4；如果 num_skips 设置为 2，则将随机选择 2 对进行训练。但是，为了训练简单，这里考虑了所有的 (target, context) 对。

为训练数据集以及验证数据集张量定义输入和输出占位符：

```
inputs = tf.placeholder(dtype=tf.int32, shape=[batch_size])
outputs = tf.placeholder(dtype=tf.int32, shape=[batch_size,1])
inputs_valid = tf.constant(x_valid, dtype=tf.int32)
```

定义一个嵌入矩阵，其行数等于词汇长度，列等于嵌入维度。该矩阵中的每一行表示词汇表中一个单词的词向量。使用在 –1.0~1.0 之间均匀采样的值填充此矩阵。

```
# 使用vocab_len行和embedding_size列定义嵌入矩阵
# 每一行代表向量表示或在vocbulary中嵌入的一个单词

embed_dist = tf.random_uniform(shape=[ptb.vocab_len, embedding_size],
                               minval=-1.0,maxval=1.0)
embed_matrix = tf.Variable(embed_dist,name='embed_matrix')
```

使用此矩阵来定义一个嵌入查找表，并通过 tf.nn.embedding_lookup( ) 来实现。tf.nn.embedding_lookup( ) 有两个参数：嵌入矩阵和输入占位符。lookup 函数返回 inputs 占位符中单词的词向量。

```
# 定义嵌入查找表
# 在输入张量中提供单词id的嵌入
embed_ltable = tf.nn.embedding_lookup(embed_matrix, inputs)
```

embed_ltable 也可以解释为输入层顶部的嵌入层。接下来，将嵌入层的输出传递到 softmax 或噪声对比估计（Noise-Contrastive Estimation，NCE）层。NCE 采用训练 Logistic 回归分类器的思想：从真实和噪声数据中学习参数。

> TensorFlow 文档更详细地描述了 NCE：
> https:// www.tensorflow.org/tutorials/word2vec.

总之，损失函数为 softmax 模型会有较高的计算代价，因为需要在整个词汇表中计算概率分布并对其进行归一化处理。基于 NCE 损失模型将其转换为二分类问题，即从噪声样本中识别真实样本。

NCE 的数学基础可以在以下 NIPS 论文中找到：Andriy Mnih 和 Koray Kavukcuoglu，Learning word embeddings efficiently with noise-contrastive estimation。该论文可从以下链接获得：http://papers.nips.cc/paper/5165-learning-word-embeddings-efficient-with-noise-contrastive-estimation.PDF。

tf.nn.nce_loss( ) 函数在计算损失值时会自动生成负样本：参数 num_sampledis 被设置为负样本数 (n_negative_samples)，此参数指定要抽取的负样本数。

```
# 定义NCE损失函数层
nce_dist = tf.truncated_normal(shape=[ptb.vocab_len, embedding_size],
                               stddev=1.0 /
                               tf.sqrt(embedding_size * 1.0)
                               )
nce_w = tf.Variable(nce_dist)
nce_b = tf.Variable(tf.zeros(shape=[ptb.vocab_len]))

loss = tf.reduce_mean(tf.nn.nce_loss(weights=nce_w,
                                     biases=nce_b,
                                     inputs=embed_ltable,
                                     labels=outputs,
                                     num_sampled=n_negative_samples,
                                     num_classes=ptb.vocab_len
                                     )
                      )
```

接下来，计算验证集中的样本与嵌入矩阵之间的余弦相似度：

1）为了计算相似性得分，首先计算嵌入矩阵中每个词向量的 L2 范数。

```
# 计算验证采样本和所有嵌入之间的余弦相似度
norm = tf.sqrt(tf.reduce_sum(tf.square(embed_matrix), 1,
                             keep_dims=True))
normalized_embeddings = embed_matrix / norm
```

2）在验证集中查找样本的词向量：

```
embed_valid = tf.nn.embedding_lookup(normalized_embeddings,
                                     inputs_valid)
```

3）通过将验证集的词向量与嵌入矩阵相乘来计算相似性分数。

```
similarity = tf.matmul(
    embed_valid, normalized_embeddings, transpose_b=True)
```

这里给出了具有 (valid_size, vocab_len) 形状的张量。张量中的每一行指的是验证单词和词汇单词之间的相似性得分。

接下来，定义 SGD 优化器，学习率为 0.9，用 50 次迭代。

```
n_epochs = 10
learning_rate = 0.9
n_batches = ptb.n_batches(batch_size)
optimizer = tf.train.GradientDescentOptimizer(learning_rate)
            .minimize(loss)
```

对于每次迭代:

1)在整个数据集上逐批运行优化程序。

```
ptb.reset_index_in_epoch()
for step in range(n_batches):
    x_batch, y_batch = ptb.next_batch()
    y_batch = dsu.to2d(y_batch,unit_axis=1)
    feed_dict = {inputs: x_batch, outputs: y_batch}
    _, batch_loss = tfs.run([optimizer, loss], feed_dict=feed_dict)
    epoch_loss += batch_loss
```

2)计算并打印每次迭代的平均损失。

```
 epoch_loss = epoch_loss / n_batches
 print('\n','Average loss after epoch ', epoch, ': ', epoch_loss)
```

3)在一次迭代结束时,计算相似性得分。

```
similarity_scores = tfs.run(similarity)
```

4)对于验证集中的每个单词,打印具有最高相似性分数的 5 个单词。

```
top_k = 5
for i in range(valid_size):
    similar_words = (-similarity_scores[i,:])
                    .argsort()[1:top_k + 1]
    similar_str = 'Similar to {0:}:'
                  .format(ptb.id2word[x_valid[i]])
    for k in range(top_k):
        similar_str = '{0:} {1:},'.format(similar_str,
                        ptb.id2word[similar_words[k]])
    print(similar_str)
```

在完成所有迭代之后,计算可在学习过程中进一步利用的嵌入向量:

```
final_embeddings = tfs.run(normalized_embeddings)
```

完整的训练代码如下:

```
n_epochs = 10
learning_rate = 0.9
n_batches = ptb.n_batches_wv()
optimizer = tf.train.GradientDescentOptimizer(learning_rate).minimize(loss)

with tf.Session() as tfs:
    tf.global_variables_initializer().run()
    for epoch in range(n_epochs):
        epoch_loss = 0
        ptb.reset_index()
        for step in range(n_batches):
            x_batch, y_batch = ptb.next_batch_sg()
```

```
            y_batch = nputil.to2d(y_batch, unit_axis=1)
            feed_dict = {inputs: x_batch, outputs: y_batch}
            _, batch_loss = tfs.run([optimizer, loss], feed_dict=feed_dict)
            epoch_loss += batch_loss
        epoch_loss = epoch_loss / n_batches
        print('\nAverage loss after epoch ', epoch, ': ', epoch_loss)

        # 在每个步长的末尾输出最接近验证集的单词
        similarity_scores = tfs.run(similarity)
        top_k = 5
        for i in range(valid_size):
            similar_words = (-similarity_scores[i, :]
                                ).argsort()[1:top_k + 1]
            similar_str = 'Similar to {0:}:'.format(
                ptb.id2word[x_valid[i]])
            for k in range(top_k):
                similar_str = '{0:} {1:},'.format(
                    similar_str, ptb.id2word[similar_words[k]])
            print(similar_str)
    final_embeddings = tfs.run(normalized_embeddings)
```

这是分别在第 1 和第 10 个步长之后获得的结果:

```
Average loss after epoch  0 :  115.644006802
Similar to we: types, downturn, internal, by, introduce,
Similar to been: said, funds, mcgraw-hill, street, have,
Similar to also: will, she, next, computer, 's,
Similar to of: was, and, milk, dollars, $,
Similar to last: be, october, acknowledging, requested, computer,
Similar to u.s.: plant, increase, many, down, recent,
Similar to an: commerce, you, some, american, a,
Similar to trading: increased, describes, state, companies, in,

Average loss after epoch  9 :  5.56538496033
Similar to we: types, downturn, introduce, internal, claims,
Similar to been: exxon, said, problem, mcgraw-hill, street,
Similar to also: will, she, ssangyong, audit, screens,
Similar to of: seasonal, dollars, motor, none, deaths,
Similar to last: acknowledging, allow, incorporated, joint, requested,
Similar to u.s.: undersecretary, typically, maxwell, recent, increase,
Similar to an: banking, officials, imbalances, americans, manager,
Similar to trading: describes, increased, owners, committee, else,
```

最后, 运行 5 000 次迭代后的模型, 并获得以下输出结果:

```
Average loss after epoch  4999 :  2.74216903135
Similar to we: matter, noted, here, classified, orders,
Similar to been: good, precedent, medium-sized, gradual, useful,
Similar to also: introduce, england, index, able, then,
Similar to of: indicator, cleveland, theory, the, load,
Similar to last: dec., office, chrysler, march, receiving,
Similar to u.s.: label, fannie, pressures, squeezed, reflection,
Similar to an: knowing, outlawed, milestones, doubled, base,
Similar to trading: associates, downturn, money, portfolios, go,
```

为了获得更好的结果，进一步运行 50 000 次迭代。

同样，在 50 次迭代之后基于 text8 得到的训练模型所输出的结果如下：

```
Average loss after epoch  49 :  5.74381046423
Similar to four: five, three, six, seven, eight,
Similar to all: many, both, some, various, these,
Similar to between: with, through, thus, among, within,
Similar to a: another, the, any, each, tpvgames,
Similar to that: which, however, although, but, when,
Similar to zero: five, three, six, eight, four,
Similar to is: was, are, has, being, busan,
Similar to no: any, only, the, another, trinomial,
```

8.4　使用 t-SNE 可视化单词嵌入

现在来可视化在上一节中生成的单词嵌入。t-SNE 是在二维空间中显示高维数据的最流行方法。这里将使用 scikit-learn 库中的方法和 TensorFlow 文档中给出的代码来绘制刚学到的嵌入词的图形。

 TensorFlow 文档中的原始代码可从以下链接获得：
https://github.com/tensorflow/tensorflow/blob/r1.3/ tensorflow/examples/tutorials/word2vec/word2vec_basic.py。

下面介绍如何实现该程序：

1）创建 tsne 模型：

```
tsne = TSNE(perplexity=30, n_components=2,
            init='pca', n_iter=5000, method='exact')
```

2）将要显示的嵌入维数限制为 500，否则，图形会变得非常难以理解：

```
n_embeddings = 500
```

3）通过调用 tsne 模型的 fit_transform( ) 方法来创建低维表示，并传递 final_embeddings 的第一个 n_embeddings 作为输入：

```
low_dim_embeddings = tsne.fit_transform(
    final_embeddings[:n_embeddings, :])
```

4）找到所选择的词向量的文本表示：

```
labels = [ptb.id2word[i] for i in range(n_embeddings)]
```

5）最后，绘制词向量示意图：

```
plot_with_labels(low_dim_embeddings, labels)
```

其结果如图 8-1 所示。

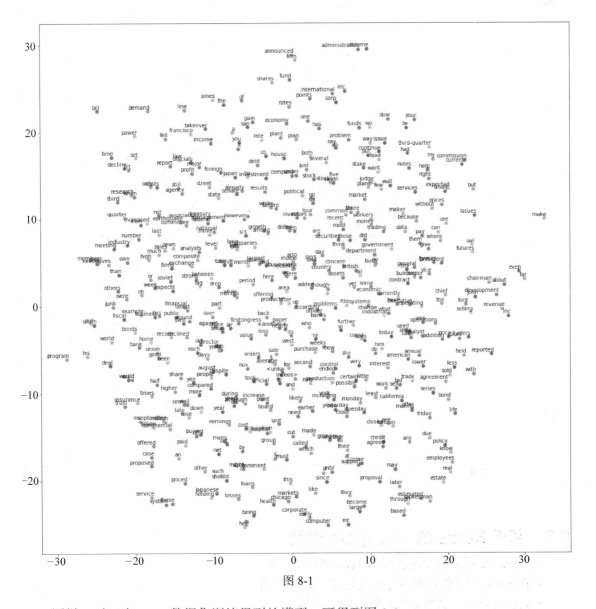

图 8-1

同样，对于由 text8 数据集训练得到的模型，可得到图 8-2。

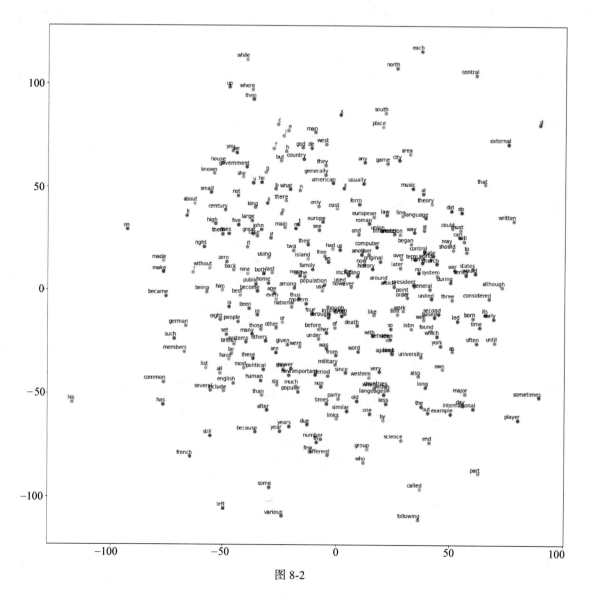

图 8-2

8.5 基于 Keras 的 skip-gram 模型

使用 Keras 嵌入模型的流程与 TensorFlow 一致。
- 在 Keras 功能性模型或序列化模型中创建网络;
- 将目标和上下文单词的真假对输入网络;
- 查找目标和上下文单词的词向量;
- 通过词向量的内积来获得相似性分数;
- 将相似性分数传给 sigmoid 层,从而得到真或假对。

现在使用 Keras 功能性 API 来实现这些步骤:

1)导入所需的库:

```
from keras.models import Model
from keras.layers.embeddings import Embedding
from keras.preprocessing import sequence
from keras.preprocessing.sequence import skipgrams
from keras.layers import Input, Dense, Reshape, Dot, merge
import keras
```

重置计算图,以便清除以前Jupyter笔记本的影响:

```
# 重置Jupyter缓冲区
tf.reset_default_graph()
keras.backend.clear_session()
```

2)创建一个验证集,用来输出模型在训练结束时找到的相似单词:

```
valid_size = 8
x_valid = np.random.choice(valid_size * 10, valid_size,
replace=False)
print('valid: ',x_valid)
```

3)定义所需的超参数:

```
batch_size = 1024
embedding_size = 512
n_negative_samples = 64
ptb.skip_window=2
```

4)使用keras.preprocessing.sequence中的make_sampling_table( )函数创建一个大小等于词汇长度的样本表。接下来,使用函数skipgrams( )从keras.preprocessing.sequence生成上下文和目标单词对,以及得到单词对是真还是假的标签。

```
sample_table = sequence.make_sampling_table(ptb.vocab_len)
pairs, labels= sequence.skipgrams(ptb.part['train'],
    ptb.vocab_len,window_size=ptb.skip_window,
    sampling_table=sample_table)
```

5)使用以下代码生成真假单词对,并输出部分单词对:

```
print('The skip-gram pairs : target,context')
for i in range(5 * ptb.skip_window):
    print(['{} {}'.format(id,ptb.id2word[id]) \
        for id in pairs[i]],':',labels[i])
```

生成的单词对如下:

```
The skip-gram pairs : target,context
['547 trying', '5 to'] : 1
['4845 bargain', '2 <eos>'] : 1
['1705 election', '198 during'] : 1
['4704 flows', '8117 gun'] : 0
['13 is', '37 company'] : 1
['625 above', '132 three'] : 1
['5768 pessimistic', '1934 immediate'] : 0
['637 china', '2 <eos>'] : 1
['258 five', '1345 pence'] : 1
['1956 chrysler', '8928 exercises'] : 0
```

6) 从上面生成的单词对中拆分目标和上下文单词, 以便能将其输入模型。将目标和上下文单词转换为二维数组。

```
x,y=zip(*pairs)
x=np.array(x,dtype=np.int32)
x=dsu.to2d(x,unit_axis=1)
y=np.array(y,dtype=np.int32)
y=dsu.to2d(y,unit_axis=1)
labels=np.array(labels,dtype=np.int32)
labels=dsu.to2d(labels,unit_axis=1)
```

7) 定义网络结构, 正如所讨论的那样, 必须将目标和上下文单词输入网络, 并且需要从嵌入层中查找向量。因此, 首先分别为目标和上下文单词定义输入层、嵌入层和重塑层:

```
# 构建目标词模型
target_in = Input(shape=(1,),name='target_in')
target = Embedding(ptb.vocab_len,embedding_size,input_length=1,
            name='target_em')(target_in)
target = Reshape((embedding_size,1),name='target_re')(target)

# 构建上下文词模型
context_in = Input((1,),name='context_in')
context = Embedding(ptb.vocab_len,embedding_size,input_length=1,
            name='context_em')(context_in)
context = Reshape((embedding_size,1),name='context_re')(context)
```

8) 接下来, 构建这两个模型的内积, 将其传递给 sigmoid 层以生成输出标签:

```
# 将两个模型内积以检查相似性并添加sigmoid层
output = Dot(axes=1,name='output_dot')([target,context])
output = Reshape((1,),name='output_re')(output)
output = Dense(1, activation='sigmoid',name='output_sig')(output)
```

9) 利用刚刚创建的输入和输出模型来构建功能模型:

```
# 创建用于查找词向量的功能性模型
model = Model(inputs=[target_in,context_in],outputs=output)
model.compile(loss='binary_crossentropy', optimizer='adam')
```

10) 此外, 在给定输入目标词的情况下, 构建一个模型, 用于预测所有单词的相似性:

```
# 合并模型并创建模型以检查余弦相似性
similarity = Dot(axes=0,normalize=True,
            name='sim_dot')([target,context])
similarity_model = Model(inputs=[target_in,context_in],
            outputs=similarity)
```

输出模型的概要信息:

```
Layer (type)                 Output Shape         Param #      Connected to
================================================================================
target_in (InputLayer)       (None, 1)            0
_____
context_in (InputLayer)      (None, 1)            0
_____
target_em (Embedding)        (None, 1, 512)       5120000      target_in[0][0]
_____
context_em (Embedding)       (None, 1, 512)       5120000      context_in[0][0]
_____
target_re (Reshape)          (None, 512, 1)       0            target_em[0][0]
_____
context_re (Reshape)         (None, 512, 1)       0            context_em[0][0]
_____
output_dot (Dot)             (None, 1, 1)         0            target_re[0][0]
                                                               context_re[0][0]
_____
output_re (Reshape)          (None, 1)            0            output_dot[0][0]
_____
output_sig (Dense)           (None, 1)            2            output_re[0][0]
================================================================================
Total params: 10,240,002
Trainable params: 10,240,002
Non-trainable params: 0
_____
```

11）接下来训练模型，训练只用了 5 次迭代，但读者可以尝试更多的迭代次数，比如 1 000 或 10 000 次迭代。

 请记住，这将需要几个小时，因为这不是最优化的代码。欢迎使用本书（或其他）的提示与技巧来进一步优化代码。

```
n_epochs = 5
batch_size = 1024
model.fit([x,y],labels,batch_size=batch_size, epochs=n_epochs)
```

根据该模型得到的词向量来计算单词的相似度：

```
# 在训练结束时输出最接近验证集的单词
top_k = 5
y_val = np.arange(ptb.vocab_len, dtype=np.int32)
y_val = dsu.to2d(y_val,unit_axis=1)
for i in range(valid_size):
    x_val = np.full(shape=(ptb.vocab_len,1),fill_value=x_valid[i],
            dtype=np.int32)
    similarity_scores = similarity_model.predict([x_val,y_val])
    similarity_scores=similarity_scores.flatten()
    similar_words = (-similarity_scores).argsort()[1:top_k + 1]
    similar_str = 'Similar to {0:}:'.format(ptb.id2word[x_valid[i]])
    for k in range(top_k):
        similar_str = '{0:} {1:},'.format(similar_str,
                    ptb.id2word[similar_words[k]])
    print(similar_str)
```

可以得到如下的输出结果：

```
Similar to we: rake, kia, sim, ssangyong, memotec,
Similar to been: nahb, sim, rake, punts, rubens,
Similar to also: photography, snack-food, rubens, nahb, ssangyong,
Similar to of: isi, rake, memotec, kia, mlx,
Similar to last: rubens, punts, memotec, sim, photography,
Similar to u.s.: mlx, memotec, punts, rubens, kia,
Similar to an: memotec, isi, ssangyong, rake, sim,
Similar to trading: rake, rubens, swapo, mlx, nahb,
```

到目前为止，已经介绍了如何使用 TensorFlow 及其高级库 Keras 创建词向量。现在来看看如何使用 TensorFlow 和 Keras 学习模型并将模型应用于一些与 NLP 相关的任务中。

8.6 使用 TensorFlow 和 Keras 中的 RNN 模型生成文本

生成文本是 RNN 模型在 NLP 中的主要应用之一，针对文本序列训练 RNN 模型，然后通过提供种子文本（seed text）作为输入生成文本序列。下面会基于 text8 数据集进行介绍。

加载 text8 数据集并输出前 100 个单词：

```
from datasetslib.text8 import Text8
text8 = Text8()
#    加载数据，将单词转换为ids，将文件转换为ids列表
text8.load_data()
print(' '.join([text8.id2word[x_i] for x_i in text8.part['train'][0:100]]))
```

得到如下输出结果：

```
anarchism originated as a term of abuse first used against early working
class radicals including the diggers of the english revolution and the sans
culottes of the french revolution whilst the term is still used in a
pejorative way to describe any act that used violent means to destroy the
organization of society it has also been taken up as a positive label by
self defined anarchists the word anarchism is derived from the greek
without archons ruler chief king anarchism as a political philosophy is the
belief that rulers are unnecessary and should be abolished although there
are differing
```

在这个示例中，将数据加载并剪切为具有 5 000 单词的文本，因为较大的文本需要更高级的技术，例如分布式或批处理技术，而这里希望示例变得简单。

```
from datasetslib.text8 import Text8
text8 = Text8()
text8.load_data(clip_at=5000)
print('Train:', text8.part['train'][0:5])
print('Vocabulary Length = ',text8.vocab_len)
```

现在单词量减少到 1 457 个。

```
Train: [ 8 497   7   5 116]
Vocabulary Length =   1457
```

在示例中，构造了一个单层 LSTM 网络。为了训练模型，使用 5 个单词作为输入，以便能学习第六个单词的参数。输入层是 5 个词，隐藏层是具有 128 个单元的 LSTM 网络，

最后一层是全连接的层，其输出等于词汇量大小。这个示例没有使用词向量，而是对输出向量采用非常简单的独热编码。

在训练完模型后，为了测试模型，可通过 2 个不同的字符串作为种子生成更多的字符：
- random5：随机选择 5 个单词生成字符串。
- first5：使用文本的前 5 个单词生成字符串。

```
random5 = np.random.choice(n_x * 50, n_x, replace=False)
print('Random 5 words: ',id2string(random5))
first5 = text8.part['train'][0:n_x].copy()
print('First 5 words: ',id2string(first5))
```

种子字符串是：

```
Random 5 words:  free bolshevik be n another
First 5 words:  anarchism originated as a term
```

不同读者执行上述代码，其随机种子字符串可能不同。下面先用 TensorFlow 创建 LSTM 模型。

8.6.1 使用 TensorFlow 中的 LSTM 模型生成文本

 可以在 Jupyter 笔记本 ch-08b_RNN_Text_TensorFlow 中按照此部分的代码进行操作。

使用以下步骤在 TensorFlow 中实现生成文本的 LSTM 网络：

1）定义 x 和 y 的参数和占位符：

```
batch_size = 128
n_x = 5 # 输入单词的数量
n_y = 1 # 输出单词的数量
n_x_vars = 1 # 在我们的文本中，每个时间步长只有一个变量
n_y_vars = text8.vocab_len
state_size = 128
learning_rate = 0.001
x_p = tf.placeholder(tf.float32, [None, n_x, n_x_vars], name='x_p')
y_p = tf.placeholder(tf.float32, [None, n_y_vars], name='y_p')
```

对于输入，使用整数表示单词，因此 n_x_vars 为 1；对于输出，使用独热编码值，因此输出的数量等于词汇长度。

2）接下来，创建长度为 n_x 的张量列表：

```
x_in = tf.unstack(x_p,axis=1,name='x_in')
```

3）然后，从输入和单元中创建 LSTM 单元和静态 RNN：

```
cell = tf.nn.rnn_cell.LSTMCell(state_size)
rnn_outputs, final_states = tf.nn.static_rnn(cell,
x_in,dtype=tf.float32)
```

4）定义最终网络层的权重、偏差和公式。最后一层只需要为第六个单词选择输出，因

此可用下面的代码来获取最后一个输出：

```
# 输出节点参数
w = tf.get_variable('w', [state_size, n_y_vars], initializer=
tf.random_normal_initializer)
b = tf.get_variable('b', [n_y_vars],
initializer=tf.constant_initializer(0.0))
y_out = tf.matmul(rnn_outputs[-1], w) + b
```

5）创建损失函数和优化器：

```
loss = tf.reduce_mean(tf.nn.softmax_cross_entropy_with_logits(
        logits=y_out, labels=y_p))
optimizer = tf.train.AdamOptimizer(learning_rate=learning_rate)
        .minimize(loss)
```

6）创建可以在会话块中运行的精度函数，以检查训练模式的精确性：

```
n_correct_pred = tf.equal(tf.argmax(y_out,1), tf.argmax(y_p,1))
accuracy = tf.reduce_mean(tf.cast(n_correct_pred, tf.float32))
```

7）最后，通过迭代 1 000 次来训练模型，并且每隔 100 次迭代就输出一次结果。此外，每隔 100 次迭代，就从上面描述的种子串输出生成的文本。

LSTM 网络和 RNN 需要在有大量迭代周期的大型数据集上训练，才能获得更好的结果。请尝试加载完整的数据集并运行迭代 50 000 或 80 000 次，同时可尝试使用其他超参数改善结果。

```
n_epochs = 1000
learning_rate = 0.001
text8.reset_index_in_epoch()
n_batches = text8.n_batches_seq(batch_size=batch_size,n_tx=n_x,n_ty=n_y)
n_epochs_display = 100

with tf.Session() as tfs:
    tf.global_variables_initializer().run()

    for epoch in range(n_epochs):
        epoch_loss = 0
        epoch_accuracy = 0
        for step in range(n_batches):
            x_batch, y_batch = text8.next_batch_seq(batch_size=batch_size,
                                n_tx=n_x,n_ty=n_y)
            y_batch = dsu.to2d(y_batch,unit_axis=1)
            y_onehot = np.zeros(shape=[batch_size,text8.vocab_len],
                        dtype=np.float32)
            for i in range(batch_size):
                y_onehot[i,y_batch[i]]=1

            feed_dict = {x_p: x_batch.reshape(-1, n_x, n_x_vars),
                        y_p: y_onehot}
```

```
            _, batch_accuracy, batch_loss = tfs.run([optimizer,accuracy,
                                    loss],feed_dict=feed_dict)
            epoch_loss += batch_loss
            epoch_accuracy += batch_accuracy

    if (epoch+1) % (n_epochs_display) == 0:
        epoch_loss = epoch_loss / n_batches
        epoch_accuracy = epoch_accuracy / n_batches
        print('\nEpoch {0:}, Average loss:{1:}, Average accuracy:{2:}'.
                format(epoch,epoch_loss,epoch_accuracy ))

        y_pred_r5 = np.empty([10])
        y_pred_f5 = np.empty([10])
        x_test_r5 = random5.copy()
        x_test_f5 = first5.copy()
        # 提供5个单词后生成10个单词的文本
        for i in range(10):
            for x,y in zip([x_test_r5,x_test_f5],
                            [y_pred_r5,y_pred_f5]):
                x_input = x.copy()
                feed_dict = {x_p: x_input.reshape(-1, n_x, n_x_vars)}
                y_pred = tfs.run(y_out, feed_dict=feed_dict)
                y_pred_id = int(tf.argmax(y_pred, 1).eval())
                y[i]=y_pred_id
                x[:-1] = x[1:]
                x[-1] = y_pred_id
        print(' Random 5 prediction:',id2string(y_pred_r5))
        print(' First 5 prediction:',id2string(y_pred_f5))
```

输出结果如下：

```
Epoch 99, Average loss:1.3972469369570415, Average
accuracy:0.8489583333333334
   Random 5 prediction: labor warren together strongly profits strongly
supported supported co without
   First 5 prediction: market own self free together strongly profits
strongly supported supported

Epoch 199, Average loss:0.7894854595263799, Average
accuracy:0.9186197916666666
   Random 5 prediction: syndicalists spanish class movements also also
anarcho anarcho anarchist was
   First 5 prediction: five civil association class movements also anarcho
anarcho anarcho anarcho

Epoch 299, Average loss:1.360412875811259, Average accuracy:0.865234375
   Random 5 prediction: anarchistic beginnings influenced true tolstoy
tolstoy tolstoy tolstoy tolstoy tolstoy
   First 5 prediction: early civil movement be for was two most most most

Epoch 399, Average loss:1.1692512730757396, Average
accuracy:0.8645833333333334
   Random 5 prediction: including war than than revolutionary than than war
```

than than
    First 5 prediction: left including including including other other other other other other

Epoch 499, Average loss:0.5921860883633295, Average accuracy:0.923828125
    Random 5 prediction: ever edited interested interested variety variety variety variety variety variety
    First 5 prediction: english market herbert strongly price interested variety variety variety variety

Epoch 599, Average loss:0.8356450994809469, Average accuracy:0.8958333333333334
    Random 5 prediction: management allow trabajo trabajo national national mag mag ricardo ricardo
    First 5 prediction: spain prior am working n war war war self self

Epoch 699, Average loss:0.7057955612738928, Average accuracy:0.8971354166666666
    Random 5 prediction: teachings can directive tend resist obey christianity author christianity christianity
    First 5 prediction: early early called social called social social social social social

Epoch 799, Average loss:0.772875706354777, Average accuracy:0.90234375
    Random 5 prediction: associated war than revolutionary revolutionary revolutionary than than revolutionary revolutionary
    First 5 prediction: political been hierarchy war than see anti anti anti anti

Epoch 899, Average loss:0.43675946692625683, Average accuracy:0.9375
    Random 5 prediction: individualist which which individualist warren warren tucker benjamin how tucker
    First 5 prediction: four at warren individualist warren published considered considered considered considered

Epoch 999, Average loss:0.23202441136042276, Average accuracy:0.9602864583333334
    Random 5 prediction: allow allow trabajo you you you you you you you
    First 5 prediction: labour spanish they they they movement movement anarcho anarcho two

生成的文本经常会出现重复的单词，若更好地训练模型，就不会出现这种情况。虽然模型的准确性可以提高到 96%，但仍然不足以生成清晰易读的文本。尝试增加 LSTM 单元/隐藏层的数量，同时在较大的数据集上以大量的迭代运行模型。

现在在 Keras 中构建相同的模型。

8.6.2　使用 Keras 中的 LSTM 模型生成文本

 可以在 Jupyter 笔记本 ch-08b_RNN_Text_Keras 中按照本节的代码进行操作。

按以下步骤使用 Keras 的 LSTM 模型生成文本：

1）首先，将所有数据转换为两个张量，因为一次输入有 5 个单词，所以张量 x 有 5 列，另外，张量 y 是只有 1 列的输出。将 y 或标签张量转换为独热编码表示。

请记住，在大型数据集的实践中，将使用 word2vec 嵌入而不是独热编码表示。

```
# 获得数据
x_train, y_train =
text8.seq_to_xy(seq=text8.part['train'],n_tx=n_x,n_ty=n_y)
# 重塑输入为 [samples, time steps, features]
x_train = x_train.reshape(x_train.shape[0], x_train.shape[1],1)
y_onehot =
np.zeros(shape=[y_train.shape[0],text8.vocab_len],dtype=np.float32)
for i in range(y_train.shape[0]):
    y_onehot[i,y_train[i]]=1
```

2）接下来，仅使用具有一个隐藏层的 LSTM 模型。由于输出不是序列，还需将 return_sequences 设置为 False：

```
n_epochs = 1000
batch_size=128
state_size=128
n_epochs_display=100

# 创建和训练模型
model = Sequential()
model.add(LSTM(units=state_size,
               input_shape=(x_train.shape[1], x_train.shape[2]),
               return_sequences=False
               )
          )
model.add(Dense(text8.vocab_len))
model.add(Activation('softmax'))
model.compile(loss='categorical_crossentropy', optimizer='adam')
model.summary()
```

该模型的信息如下：

```
Layer (type)                 Output Shape              Param #
=================================================================
lstm_1 (LSTM)                (None, 128)               66560
_____
dense_1 (Dense)              (None, 1457)              187953
_____
activation_1 (Activation)    (None, 1457)              0
=================================================================
Total params: 254,513
Trainable params: 254,513
Non-trainable params: 0
_____
```

3）基于 Keras 实现的 LSTM 模型只需运行 10 次迭代，每次迭代用 100 个周期训练模型，并输出文本生成的结果。以下是训练模型和生成文本的完整代码：

```
for j in range(n_epochs // n_epochs_display):
    model.fit(x_train, y_onehot, epochs=n_epochs_display,
              batch_size=batch_size,verbose=0)
    # 生成文本
    y_pred_r5 = np.empty([10])
    y_pred_f5 = np.empty([10])
    x_test_r5 = random5.copy()
    x_test_f5 = first5.copy()
    # 提供5个单词后生成10个单词的文本
    for i in range(10):
        for x,y in zip([x_test_r5,x_test_f5],
                       [y_pred_r5,y_pred_f5]):
            x_input = x.copy()
            x_input = x_input.reshape(-1, n_x, n_x_vars)
            y_pred = model.predict(x_input)[0]
            y_pred_id = np.argmax(y_pred)
            y[i]=y_pred_id
            x[:-1] = x[1:]
            x[-1] = y_pred_id
    print('Epoch: ',((j+1) * n_epochs_display)-1)
    print(' Random5 prediction:',id2string(y_pred_r5))
    print(' First5 prediction:',id2string(y_pred_f5))
```

4）输出结果并不令人吃惊，从重复单词的角度来看，模型有所改进，可以通过更多的LSTM层、更多数据、更多训练迭代和其他超参数调整来进一步改进模型。

```
Random 5 words: free bolshevik be n another
First 5 words: anarchism originated as a term
```

预测的输出结果如下：

```
Epoch: 99
    Random5 prediction: anarchistic anarchistic wrote wrote wrote
wrote wrote wrote wrote wrote
    First5 prediction: right philosophy than than than than than
than than than

Epoch: 199
    Random5 prediction: anarchistic anarchistic wrote wrote wrote
wrote wrote wrote wrote wrote
    First5 prediction: term i revolutionary than war war french
french french french

Epoch: 299
    Random5 prediction: anarchistic anarchistic wrote wrote wrote
wrote wrote wrote wrote wrote
    First5 prediction: term i revolutionary revolutionary
revolutionary revolutionary revolutionary revolutionary
revolutionary revolutionary

Epoch: 399
    Random5 prediction: anarchistic anarchistic wrote wrote wrote
```

```
wrote wrote wrote wrote wrote
    First5 prediction: term i revolutionary labor had had french
french french french

Epoch: 499
    Random5 prediction: anarchistic anarchistic amongst wrote wrote
wrote wrote wrote wrote wrote
    First5 prediction: term i revolutionary labor individualist had
had french french french

Epoch: 599
    Random5 prediction: tolstoy wrote tolstoy wrote wrote wrote
wrote wrote wrote wrote      First5 prediction: term i revolutionary
labor individualist had had had had had

Epoch: 699
    Random5 prediction: tolstoy wrote tolstoy wrote wrote wrote
wrote wrote wrote wrote      First5 prediction: term i revolutionary
labor individualist had had had had had

Epoch: 799
    Random5 prediction: tolstoy wrote tolstoy tolstoy tolstoy
tolstoy tolstoy tolstoy tolstoy tolstoy
    First5 prediction: term i revolutionary labor individualist had
had had had had

Epoch: 899
    Random5 prediction: tolstoy wrote tolstoy tolstoy tolstoy
tolstoy tolstoy tolstoy tolstoy tolstoy
    First5 prediction: term i revolutionary labor should warren
warren warren warren warren

Epoch: 999
    Random5 prediction: tolstoy wrote tolstoy tolstoy tolstoy
tolstoy tolstoy tolstoy tolstoy tolstoy
    First5 prediction: term i individualist labor should warren
warren warren warren warren
```

> 读者是否注意到在 LSTM 模型的输出中有重复的单词用于文本生成。虽然超参数和网络调整可以消除一些重复，但还有其他方法可以解决这个问题。得到重复单词的原因是模型总是从单词的概率分布中选择具有最高概率的单词。可在连续单词之间引入更大可变性来进一步改进该模型。

8.7 总结

为了找到更好的文本数据的表示，本章引入了单词嵌入方法。由于神经网络和深度学习需要大量的文本数据，因此独热编码表示和其他单词表示方法不再有效。本章介绍了如何使用 t-SNE 图来可视化文字嵌入；还介绍使用 TensorFlow 和 Keras 中的简单 LSTM 模型来生成文本。类似的概念可以应用于其他任务，例如情绪分析、问答系统和神经机器翻译。

在深入研究先进的 TensorFlow 功能（如迁移学习、强化学习、生成网络和分布式 TensorFlow）之前，将在下一章介绍如何在生产环境中使用 TensorFlow 模型。

第 9 章 基于 TensorFlow 和 Keras 的 CNN

卷积神经网络（CNN，或 ConvNets）是一种特殊的前馈神经网络，在其结构中包含卷积层和池化层。通常 CNN 的层次结构为：

1）全连接的输入层；
2）多个卷积、池化和全连接层的组合；
3）全连接的输出层与 softmax 函数。

CNN 在解决图像的问题（例如图像识别和对象识别）方面非常成功。

本章将介绍 CNN 相关的内容：

- 理解卷积。
- 理解池化。
- CNN 架构模式 –LeNet。
- 在 MNIST 数据集上构建 LeNet：
  - 使用 TensorFlow 的 LeNet CNN 对 MNIST 数据集进行分类；
  - 使用 Keras 的 LeNet CNN 对 MNIST 数据集进行分类。
- 在 CIFAR 数据集上构建 LeNet：
  - 使用 TensorFlow 的 CNN 对 CIFAR10 数据集进行分类；
  - 使用 Keras 的 CNN 对 CIFAR10 数据集进行分类。

下面介绍 CNN 的核心概念。

9.1 理解卷积

卷积是 CNN 的核心概念，简单来说，卷积是一种数学运算，它将两个来源的信息结合起来产生一组新的信息。具体来说，将一个称为核的特殊矩阵应用于输入张量上，以生成一组称为特征图的矩阵。可以使用任何流行的算法将核应用于输入张量。

生成卷积矩阵的最常用算法如下：

```
N_STRIDES = [1,1]
1. Overlap the kernel with the top-left cells of the image matrix.
2. Repeat while the kernel overlaps the image matrix:
    2.1 c_col = 0
```

```
2.2 Repeat while the kernel overlaps the image matrix:
    2.1.1 set c_row = 0
    2.1.2 convolved_scalar = scalar_prod(kernel, overlapped cells)
    2.1.3 convolved_matrix(c_row,c_col) = convolved_scalar
    2.1.4 Slide the kernel down by N_STRIDES[0] rows.
    2.1.5 c_row = c_row  + 1
2.3 Slide the kernel to (topmost row, N_STRIDES[1] columns right)
2.4 c_col = c_col + 1
```

假设核矩阵是 2×2 矩阵，输入图像是 3×3 矩阵。图 9-1 给出了上述算法的执行过程。

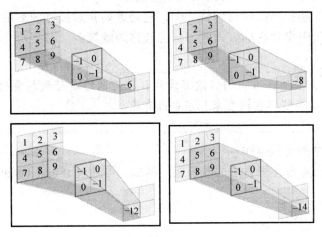

图 9-1

在执行完卷积操作后，会得到图 9-2 所示的特征图。

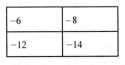

图 9-2

在上面的示例中，与卷积的原始输入相比，得到的特征图会变得更小。通常，经过卷积所得到的特征图的大小等于（核大小 −1）。因此，特征图的大小为：

$$\text{size}_{\text{feature\_map}} = \text{size}_{\text{feature}} - \text{size}_{\text{kernel}} + 1$$

1. 三维张量

对于具有深度维度的三维张量，可以考虑将前面的算法应于深度维度中的每一层中。将卷积应用于三维张量后，所得到的输出也是二维张量，因为卷积操作会与三个通道相加。

2. 步长

数组 N_STRIDES 中的步长是核跨过的行或列的大小。在这里的示例中，所使用的步长为 1。如果使用更大的步长，则特征图的大小会进一步减小，计算特征图大小的公式为：

$$\text{size}_{\text{feature\_map}} = \frac{\text{size}_{\text{features}} - \text{size}_{\text{kernel}}}{n_{\text{strides}}} + 1$$

3. 填充（padding）

如果不希望减小特征图的大小，可以填充输入数据的所有边，使得特征的大小增加。在使用填充后，可以按下式计算特征图的大小：

$$size_{feature\_map} = \frac{size_{features} + 2 \times size_{padding} - size_{kernel}}{n_{strides}} + 1$$

TensorFlow 允许两种填充：SAME 或 VALID。SAME 表示通过填充使输出的特征图与输入特征具有相同的大小。VALID 表示不进行填充。

采用前面介绍的卷积算法会得到特征图，它是原始张量被滤波后的版本。例如，特征图可以仅从原始图像中得到轮廓。因此，核也被称为滤波器。对于每个核，将获得单独的二维特征图。

用户需要选择恰当的滤波器才能让神经网络学习到想要的特征。使用 CNN 则可以自动了解哪些核在卷积层中最有效。

4. TensorFlow 中的卷积操作

TensorFlow 提供实现卷积算法的卷积层，例如，可通过 tf.nn.conv2d() 构建卷积层，该函数的定义为：

```
tf.nn.conv2d(
  input,
  filter,
  strides,
  padding,
  use_cudnn_on_gpu=None,
  data_format=None,
  name=None
)
```

input 和 filter 表示数据张量的形状 [batch_size，input_height，input_width, input_depth] 和核张量的形状 [filter_height, filter_width，input_depth, output_depth]。核张量中的 output_depth 参数表示输入的核的数量。步长张量表示每个维度中要滑动的单元格数量。参数 padding 可取两种值：VALID 和 SAME。

可以在以下链接找到有关 TensorFlow 中卷积操作的更多信息：https://www.tensorflow.org/api_guides/python/nn#Convolution。

可以在以下链接中找到有关 Keras 中卷积层的更多信息：https://keras.io/layers/convolutional/。

以下链接给出了卷积的数学解释：http://colah.github.io/posts/2014-07-Understanding-Convolutions/；http://ufldl.stanford.edu/tutorial/supervised/ FeatureExtractionUsingConvolution/。

卷积层将输入值与下一个隐藏层神经元相连。每个隐藏层神经元与输入神经元相连接，输入神经元的数量与核的大小一样。所以，在前面的例子中，核的大小为 4，因此隐藏层神经元连接到输入层的 4 个神经元（在 3×3 个神经元中）。在这个示例中，输入层的 4 个神经元所在的区域被称为**感受野**（receptive field）。

卷积层的每个核有单独权重参数和偏差参数。权重参数的数量等于核中元素的数量，而偏差参数只有一个。核的所有连接共享相同的权重参数和偏置参数。因此，在这个示例中，将有 4 个权重参数和 1 个偏差参数，但如果在卷积层中使用 5 个核，则总共将有 5×4 个权重参数和 5×1 个偏差参数，每个特征图有一组（4 个权重，1 个偏差）参数。

9.2 理解池化

通常在卷积操作中会采用几种不同的核，这些核会生成若干特征图。因此，卷积运算会生成较大的数据集。

例如，将大小为 3×3×1 的核应用于图像像素为 28×28×1 的 MNIST 数据集中，可生成大小为 26×26×1 的特征图。如果将 32 个这样的滤波器应用于卷积层，则输出的大小为 32×26×26×1，即大小为 26×26×1 的特征图有 32 个。

与形状为 28×28×1 的原始数据集相比，这是一个较大的数据集。因此，为了简化下一层的学习，人们引入池化技术。

池化是对卷积特征空间区域的聚合统计。两个最常用的池化方法是最大池化和平均池化。最大池化会取所选区域的最大值，而平均池化会计算所选区域的平均值。

例如，假设特征图的形状为 3×3，池化区域为 2×2。图 9-3 展示了基于 [1,1] 步长的最大池化操作。

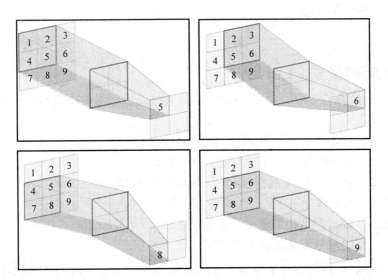

图 9-3

在完成最大池化之后，得到图 9-4 所示矩阵。

5	6
8	9

图 9-4

通常，池化操作会应用于非重叠区域，因此步长张量和区域张量被设置为相同的值。例如，TensorFlow 支持的最大池化操作的定义如下：

```
max_pool(
  value,
  ksize,
  strides,
  padding,
  data_format='NHWC',
  name=None
)
```

value 表示形状 [batch_size, input_height, input_width, input_depth] 的输入张量。对形状为 ksize 的矩形区域执行池化操作，这些区域会按 strides 参数值进行移动。

可以在以下链接找到有关 TensorFlow 中有用的池化操作信息：https://www.tensorflow.org/api_guides/python/nn#Pooling。

可以在以下链接中找到有关 Keras 中有用的池化层的更多信息：https://keras.io/layers/pooling/。

以下链接给出了池化操作的数学解释：http://ufldl.stanford.edu/tutorial/supervised/Pooling/。

9.3 CNN 架构模式 - LeNet

LeNet 是一种经典的 CNN。本章将介绍如何通过以下步骤来构建 LeNet 的层：

1）输入层；
2）卷积层 1，它会产生一组特征图，激活函数为 ReLU；
3）池化层 1，它会产生一组经过池化的特征图；
4）卷积层 2，它会产生一组特征图，激活函数为 ReLU；
5）池化层 2，它会产生一组经过池化的特征图；
6）全连接层，通过激活函数 ReLU 来展平（flatten）特征图；
7）输出层，通过线性激活函数来产生输出。

Yann LeCun 等首先提出了 LeNet 模型。有关 LeNet 的更多详细信息，请访问以下链接：http://yann.lecun.com/exdb/publis/pdf/lecun-01a.pdf。

以下链接是 Yann LeCun 维护的 LeNet 模型：http://yann.lecun.com/exdb/lenet/index.html。

9.4 在 MNIST 数据集上构建 LeNet

 可以按照 Jupyter 笔记本 ch-09a_CNN_MNIST_TF_and_Keras 中的代码进行操作。

将 MNIST 数据集分为测试数据集和训练数据集：

```
from tensorflow.examples.tutorials.mnist import input_data
mnist = input_data.read_data_sets(os.path.join('.','mnist'), one_hot=True)
X_train = mnist.train.images
X_test = mnist.test.images
Y_train = mnist.train.labels
Y_test = mnist.test.labels
```

9.4.1 使用 TensorFlow 的 LeNet CNN 对 MNIST 数据集进行分类

在 TensorFlow 中，可按以下步骤为 MNIST 数据集构建 LeNet 模型：

1）定义超参数，以及 x 和 y 的占位符（输入图像和输出标签）：

```
n_classes = 10 # 0~9 位数
n_width = 28
n_height = 28
n_depth = 1
n_inputs = n_height * n_width * n_depth # 总像素
learning_rate = 0.001
n_epochs = 10
batch_size = 100
n_batches = int(mnist.train.num_examples/batch_size)
# 输入图像形状(n_samples,n_pixels)
x = tf.placeholder(dtype=tf.float32, name="x", shape=[None,
n_inputs])
# 输出标签
y = tf.placeholder(dtype=tf.float32, name="y", shape=[None,
n_classes])
```

转换输入 x 为形状 (n_samples, n_width, n_height, n_depth)：

```
x_=tf.reshape(x, shape=[-1, n_width, n_height, n_depth])
```

2）使用 32 个 4×4 大小的核定义第一个卷积层，从而生成 32 个特征图。

- 首先，定义第一个卷积层的权重和偏差，使用正态分布初始化这些参数：

```
layer1_w = tf.Variable(tf.random_normal(shape=[4,4,n_depth,32],
            stddev=0.1),name='l1_w')
layer1_b = tf.Variable(tf.random_normal([32]),name='l1_b')
```

- 接下来，使用 tf.nn.conv2d 函数定义卷积层。函数参数 stride 定义了核张量在每个维度中所滑动的元素的数量。维度顺序由参数 data_format 决定，可以是 'NHWC' 或 'NCHW'（默认情况下为 'NHWC'）。通常，步长列表的第一个元素和最后一个元素都设置为 '1'，参数 padding 的取值为 SAME 和 VALID。SAME 表示用零填充输入数据，使得在卷积之后输出

143

与输入具有相同的大小。使用 tf.nn.relu( ) 添加 relu 激活函数：

```
layer1_conv = tf.nn.relu(tf.nn.conv2d(x_,layer1_w,
                                      strides=[1,1,1,1],
                                      padding='SAME'
                                     ) +
                         layer1_b
                        )
```

- 使用 tf.nn.max_pool() 函数定义第一个池化层。参数 ksize 表示在 $2\times2\times1$ 的区域上进行池化操作，参数 stride 表示滑动区域的大小为 $2\times2\times1$ 个像素。因此，这些区域彼此不重叠。由于使用最大池化，因此池化操作会选择 $2\times2\times1$ 区域中的最大值：

```
layer1_pool = tf.nn.max_pool(layer1_conv,ksize=[1,2,2,1],
                             strides=[1,2,2,1],padding='SAME')
```

第一个卷积层产生 32 个大小为 $28\times28\times1$ 的特征图，然后池化成 $32\times14\times14\times1$。

3）定义的第二个卷积层会将此数据作为输入，并生成 64 个特征图。
- 首先，定义第二个卷积层的权重和偏差，并使用正态分布的值初始化参数：

```
layer2_w = tf.Variable(tf.random_normal(shape=[4,4,32,64],
          stddev=0.1),name='l2_w')
layer2_b = tf.Variable(tf.random_normal([64]),name='l2_b')
```

接下来，使用 tf.nn.conv2d 函数定义卷积层：

```
layer2_conv = tf.nn.relu(tf.nn.conv2d(layer1_pool,
                                      layer2_w,
                                      strides=[1,1,1,1],
                                      padding='SAME'
                                     ) +
                         layer2_b
                        )
```

- 使用 tf.nn.max_pool 函数定义第二个池化层：

```
layer2_pool = tf.nn.max_pool(layer2_conv,
                             ksize=[1,2,2,1],
                             strides=[1,2,2,1],
                             padding='SAME'
                            )
```

第二卷积层输出的大小为 $64\times14\times14\times1$，池化之后为 $64\times7\times7\times1$。

4）在输入给全连接层（它有 1024 个神经元）之前需要将输出结果拉伸（flat）成大小为 1024 的向量：

```
layer3_w = tf.Variable(tf.random_normal(shape=[64*7*7*1,1024],
          stddev=0.1),name='l3_w')
layer3_b = tf.Variable(tf.random_normal([1024]),name='l3_b')
layer3_fc = tf.nn.relu(tf.matmul(tf.reshape(layer2_pool,
            [-1, 64*7*7*1]),layer3_w) + layer3_b)
```

5）全连接层的输出与一个线性输出层（它有 10 个输出）相连，这一层没有使用

softmax，因为损失函数会自动将 softmax 应用于输出：

```
layer4_w = tf.Variable(tf.random_normal(shape=[1024, n_classes],
                                        stddev=0.1),name='l)
layer4_b = tf.Variable(tf.random_normal([n_classes]),name='l4_b')
layer4_out = tf.matmul(layer3_fc,layer4_w)+layer4_b
```

这是创建的第一个 CNN 模型，它保存在变量 model 中：

```
model = layer4_out
```

 读者可以通过取不同的超参数值来研究一下 TensorFlow 中的其他卷积操作符和池化操作符。

可用 tf.nn.softmax_cross_entropy_with_logits 函数定义损失函数。使用 AdamOptimizer 函数作为优化器。读者可以尝试研究 TensorFlow 中其他可用的优化器函数。

```
entropy = tf.nn.softmax_cross_entropy_with_logits(logits=model, labels=y)
loss = tf.reduce_mean(entropy)
optimizer = tf.train.AdamOptimizer(learning_rate).minimize(loss)
```

最后，通过遍历 n_epochs 训练模型，每个迭代周期内要训练的批次数量为 n_batches，每个批次的样本数量为 batch_size：

```
with tf.Session() as tfs:
    tf.global_variables_initializer().run()
    for epoch in range(n_epochs):
        total_loss = 0.0
        for batch in range(n_batches):
            batch_x,batch_y = mnist.train.next_batch(batch_size)
            feed_dict={x:batch_x, y: batch_y}
            batch_loss,_ = tfs.run([loss, optimizer],
                                   feed_dict=feed_dict)
            total_loss += batch_loss
        average_loss = total_loss / n_batches
        print("Epoch: {0:04d} loss = {1:0.6f}".format(epoch,average_loss))
    print("Model Trained.")

    predictions_check = tf.equal(tf.argmax(model,1),tf.argmax(y,1))
    accuracy = tf.reduce_mean(tf.cast(predictions_check, tf.float32))
    feed_dict = {x:mnist.test.images, y:mnist.test.labels}
    print("Accuracy:", accuracy.eval(feed_dict=feed_dict))
```

得到以下输出结果：

```
Epoch: 0000    loss = 1.418295
Epoch: 0001    loss = 0.088259
Epoch: 0002    loss = 0.055410
Epoch: 0003    loss = 0.042798
Epoch: 0004    loss = 0.030471
Epoch: 0005    loss = 0.023837
Epoch: 0006    loss = 0.019800
```

```
Epoch: 0007    loss = 0.015900
Epoch: 0008    loss = 0.012918
Epoch: 0009    loss = 0.010322
Model Trained.
Accuracy: 0.9884
```

与在前几章中所介绍的方法相比,这里的分类精度非常高。从图像数据中学习 CNN 模型是不是很神奇呢?

9.4.2 使用 Keras 的 LeNet CNN 对 MNIST 数据集进行分类

下面用相同数据集重新实现 LeNet 网络,用 Keras 构建和训练 CNN 模型的步骤为:

1)导入所需的 Keras 模块:

```
import keras
from keras.models import Sequential
from keras.layers import Conv2D,MaxPooling2D, Dense, Flatten, Reshape
from keras.optimizers import SGD
```

2)定义每个层的滤波器数量:

```
n_filters=[32,64]
```

3)定义其他超参数:

```
learning_rate = 0.01
n_epochs = 10
batch_size = 100
```

4)定义序列模型,添加一个层,并将输入数据变换为(n_width, n_height, n_depth):

```
model = Sequential()
model.add(Reshape(target_shape=(n_width,n_height,n_depth),
                input_shape=(n_inputs,))
        )
```

5)使用 4×4 大小的滤波器,填充方式为 SAME,第一个卷积层的激活函数为 relu:

```
model.add(Conv2D(filters=n_filters[0],kernel_size=4,
                padding='SAME',activation='relu')
        )
```

6)添加区域大小为 2×2,且步长为 2×2 的池化层:

```
model.add(MaxPooling2D(pool_size=(2,2),strides=(2,2)))
```

7)按添加第一个卷积和池化层的方式添加第二个卷积和池化层:

```
model.add(Conv2D(filters=n_filters[1],kernel_size=4,
                padding='SAME',activation='relu')
        )
model.add(MaxPooling2D(pool_size=(2,2),strides=(2,2)))
```

8)添加一个层来作为第二层的输出,并添加具有 1024 个神经元的全连接层,以处理展平的输出:

```
model.add(Flatten())
model.add(Dense(units=1024, activation='relu'))
```

9）将 softmax 激活函数添加到最后的输出层：

```
model.add(Dense(units=n_outputs, activation='softmax'))
```

10）使用以下代码查看模型概要：

```
model.summary()
```

模型描述如下：

```
Layer (type)                 Output Shape              Param #
=================================================================
reshape_1 (Reshape)          (None, 28, 28, 1)         0
_____
conv2d_1 (Conv2D)            (None, 28, 28, 32)        544
_____
max_pooling2d_1 (MaxPooling2 (None, 14, 14, 32)        0
_____
conv2d_2 (Conv2D)            (None, 14, 14, 64)        32832
_____
max_pooling2d_2 (MaxPooling2 (None, 7, 7, 64)          0
_____
flatten_1 (Flatten)          (None, 3136)              0
_____
dense_1 (Dense)              (None, 1024)              3212288
_____
dense_2 (Dense)              (None, 10)                10250
=================================================================
Total params: 3,255,914
Trainable params: 3,255,914
Non-trainable params: 0
_____
```

11）编译、训练和评估模型：

```
model.compile(loss='categorical_crossentropy',
              optimizer=SGD(lr=learning_rate),
              metrics=['accuracy'])
model.fit(X_train, Y_train,batch_size=batch_size,
          epochs=n_epochs)
score = model.evaluate(X_test, Y_test)
print('\nTest loss:', score[0])
print('Test accuracy:', score[1])
```

得到以下输出结果：

```
Epoch 1/10
55000/55000 [==============================] - 267s - loss: 0.8854 - acc: 0.7631
Epoch 2/10
55000/55000 [==============================] - 272s - loss: 0.2406 - acc: 0.9272
Epoch 3/10
55000/55000 [==============================] - 267s - loss: 0.1712 - acc: 0.9488
Epoch 4/10
55000/55000 [==============================] - 295s - loss: 0.1339 - acc: 0.9604
```

```
Epoch 5/10
55000/55000 [==================] - 278s - loss: 0.1112 - acc: 0.9667
Epoch 6/10
55000/55000 [==================] - 279s - loss: 0.0957 - acc: 0.9714
Epoch 7/10
55000/55000 [==================] - 316s - loss: 0.0842 - acc: 0.9744
Epoch 8/10
55000/55000 [==================] - 317s - loss: 0.0758 - acc: 0.9773
Epoch 9/10
55000/55000 [==================] - 285s - loss: 0.0693 - acc: 0.9790
Epoch 10/10
55000/55000 [==================] - 217s - loss: 0.0630 - acc: 0.9804
Test loss: 0.0628845927377
Test accuracy: 0.9785
```

这里的分类精度与前面有所不同，其原因在于训练过程中采用了 SGD 优化器，这里没有使用 TensorFlow 模型中 AdamOptimizer 所提供的一些高级功能。

9.5 在 CIFAR10 数据集上构建 LeNet

现在已经学会使用 TensorFlow 和 Keras 在 MNIST 数据集构建和训练 CNN 模型，下面在 CIFAR10 数据集上重复上面的方法。

CIFAR-10 数据集由 60 000 幅大小为 32×32 像素的 RGB 图像组成。图像被平均分为 10 个不同的类别：飞机、汽车、鸟、猫、鹿、狗、青蛙、马、船和卡车。CIFAR-10 和 CIFAR-100 是具有 8 000 万幅大图像数据集的子集。Alex Krizhevsky、Vinod Nair 和 Geoffrey Hinton 收集并标注了 CIFAR 数据集。数字 10 和 100 表示图像类别数量。

有关 CIFAR 数据集的更多详细信息，请访问以下链接：http://www.cs.toronto.edu/~kriz/cifar.html 和 http://www.cs.toronto.edu/~kriz/learning-features-2009-TR.pdf。

选择 CIFAR 10 数据集是因为它包含的图像有 3 个通道，即图像的深度为 3，而 MNIST 数据集中的图像只有一个通道。为简洁起见，省略了下载过程，以及如何将数据集拆分为训练集和测试集的介绍，在本书的 datasetslib 包中提供了这些步骤的实现代码。

可以按照 Jupyter 笔记本 ch-09b_CNN_CIFAR10_TF_and_Keras 中的代码进行操作。

使用以下代码来加载和预处理 CIFAR10 数据：

```
from datasetslib.cifar import cifar10
from datasetslib import imutil
dataset = cifar10()
dataset.x_layout=imutil.LAYOUT_NHWC
dataset.load_data()
dataset.scaleX()
```

加载数据时让图像为"NHWC"格式,这使得数据的形状变为(number_of_samples, image_height, image_width, image_channels)。图像通道称为图像深度,图像中的每个像素的取值范围是 0~255 之间的数字。使用 MinMax 归一化数据集,这种方法会将所有像素值除以 255。

加载和预处理数据之后,便得到了数据集的对象变量:dataset.X_train、dataset.Y_train、dataset.X_test 和 dataset.Y_test。

9.5.1 使用 TensorFlow 的 CNN 对 CIFAR10 数据集进行分类

保持层、滤波器以及它们的大小与之前的 MNIST 示例相同,只增加一个正则化层。由于与 MNIST 相比,这个数据集很复杂,需要进行正则化,因此添加了一个 dropout 层:

```
tf.nn.dropout(layer1_pool, keep_prob)
```

在预测和评估期间,占位符 keep_prob 设置为 1。这样就可以重复使用相同的模型进行训练、预测和评估。

基于 CIFAR10 数据集的 LeNet 模型的完整代码可在 Jupyter 笔记本 ch-09b_CNN_CIFAR10_TF_and_Keras 中找到。

运行模型得到以下的输出结果:

```
Epoch: 0000   loss = 2.115784
Epoch: 0001   loss = 1.620117
Epoch: 0002   loss = 1.417657
Epoch: 0003   loss = 1.284346
Epoch: 0004   loss = 1.164068
Epoch: 0005   loss = 1.058837
Epoch: 0006   loss = 0.953583
Epoch: 0007   loss = 0.853759
Epoch: 0008   loss = 0.758431
Epoch: 0009   loss = 0.663844
Epoch: 0010   loss = 0.574547
Epoch: 0011   loss = 0.489902
Epoch: 0012   loss = 0.410211
Epoch: 0013   loss = 0.342640
Epoch: 0014   loss = 0.280877
Epoch: 0015   loss = 0.234057
Epoch: 0016   loss = 0.195667
Epoch: 0017   loss = 0.161439
Epoch: 0018   loss = 0.140618
Epoch: 0019   loss = 0.126363
Model Trained.
Accuracy: 0.6361
```

从上面的结果可以看出:在 CIFAR10 上的分类精度没有 MNIST 上的高。通过调整不同的超参数并改变卷积和池化层的组合,可以得到更高的分类精度。读者可以尝试不同的 LeNet 架构和超参数实现更高的分类精度。

9.5.2 使用 Keras 的 CNN 对 CIFAR10 数据集进行分类

在 Keras 中，重复 LeNet 模型构建和训练 CIFAR10 数据集。保持代码架构与前面的示例相同。在 Keras 中，添加 dropout 层的方法如下：

```
model.add(Dropout(0.2))
```

Jupyter 笔记本 ch-09b_CNN_CIFAR10_TF_and_Keras 中提供了构建这一模型的完整代码。

下面是该模型的摘要：

```
Layer (type)                 Output Shape              Param #
=================================================================
conv2d_1 (Conv2D)            (None, 32, 32, 32)        1568

max_pooling2d_1 (MaxPooling2 (None, 16, 16, 32)        0

dropout_1 (Dropout)          (None, 16, 16, 32)        0

conv2d_2 (Conv2D)            (None, 16, 16, 64)        32832

max_pooling2d_2 (MaxPooling2 (None, 8, 8, 64)          0

dropout_2 (Dropout)          (None, 8, 8, 64)          0

flatten_1 (Flatten)          (None, 4096)              0

dense_1 (Dense)              (None, 1024)              4195328

dropout_3 (Dropout)          (None, 1024)              0

dense_2 (Dense)              (None, 10)                10250
=================================================================
Total params: 4,239,978
Trainable params: 4,239,978
Non-trainable params: 0
```

下面是得到的训练结果和评估结果：

```
Epoch 1/10
50000/50000 [==============================] - 191s - loss: 1.5847 - acc: 0.4364
Epoch 2/10
50000/50000 [==============================] - 202s - loss: 1.1491 - acc: 0.5973
Epoch 3/10
50000/50000 [==============================] - 223s - loss: 0.9838 - acc: 0.6582
Epoch 4/10
50000/50000 [==============================] - 223s - loss: 0.8612 - acc: 0.7009
Epoch 5/10
50000/50000 [==============================] - 224s - loss: 0.7564 - acc: 0.7394
Epoch 6/10
50000/50000 [==============================] - 217s - loss: 0.6690 - acc: 0.7710
```

```
Epoch 7/10
50000/50000 [====================] - 222s - loss: 0.5925 - acc: 0.7945
Epoch 8/10
50000/50000 [====================] - 221s - loss: 0.5263 - acc: 0.8191
Epoch 9/10
50000/50000 [====================] - 237s - loss: 0.4692 - acc: 0.8387
Epoch 10/10
50000/50000 [====================] - 230s - loss: 0.4320 - acc: 0.8528
Test loss: 0.849927025414
Test accuracy: 0.7414
```

读者可以尝试用不同的 LeNet 架构和超参数获得更高的分类精度。

9.6 总结

本章介绍了如何使用 TensorFlow 和 Keras 创建卷积神经网络（CNN），学习了卷积和池化的核心概念，这些是 CNN 的基础。另外，本章还学习了 LeNet 的结构，并在 MNIST 数据集和 CIFAR 数据集上创建、训练和评估了 LeNet 模型。TensorFlow 和 Keras 提供了许多卷积层操作和池化层操作。建议读者去研究本章未涉及的层和相关的操作。

下一章将学习自编码器（AutoEncoder）的结构，并将介绍如何在图像数据上使用 TensorFlow 中的自编码器。

第 10 章 基于 TensorFlow 和 Keras 的自编码器

自编码器是一种神经网络结构，通常与无监督学习、降维和数据压缩有关。自编码器通过使用隐藏层中较少数量的神经元学习生成与输入层相同的输出。这允许隐藏层以较少数量的参数学习输入的特征，使用较少数量的神经元学习输入数据特征会减少输入数据的维度。

自编码器的结构分为两部分：编码器和解码器。在编码器阶段，模型学会了表示输入数据，使其成为具有较小维度的压缩向量；在解码器阶段，模型将压缩向量表示为输出向量。损失函数为输出与输入之间的熵距离（entropy distance），因此通过最小化损失函数，整个学习过程会将输入数据编码成一种新的表示形式，这种形式与原来的输入尽可能相似。该过程会学习到一组参数。

本章将介绍如何使用 TensorFlow 和 Keras 按以下主题创建自编码器：

- 自编码器类型；
- 基于 TensorFlow 和 Keras 的堆叠自编码器；
- 基于 TensorFlow 和 Keras 的去噪自编码器；
- 基于 TensorFlow 和 Keras 的变分自编码器。

10.1 自编码器类型

有多种自编码器的结构，例如简单的自编码器、稀疏自编码器、去噪自编码器和卷积自编码器。

- **简单的自编码器**：在简单的自编码器中，与输入相比，隐藏层具有较少数量的节点或神经元。例如，在 MNIST 数据集中，输入数据有 784 个特征，与之连接的隐藏层的节点数为 512 或 256，该隐藏层又与输出层（有 784 个神经元）连接。因此，在训练期间，仅由 256 个节点学习 784 个特征。简单的自编码器也称为欠完备（undercomplete）自编码器。

简单的自编码器可以是单层或多层。通常，单层自编码器在实际应用中表现不好。多层自编码器具有多个隐藏层，分为编码器和解码器。编码器层将大量特征编码为较少的神经元，然后解码器层将学习的压缩特征解码成跟原来一样的特征数量。多层自编码器称为堆叠自编码器。

- **稀疏自编码器**：在稀疏自编码器中，添加正则化项作为惩罚，因此，与简单自编码器相比，会对数据的表示变得更稀疏。
- **去噪自编码器（DAE）**：在 DAE 架构中，输入带有随机噪声。DAE 重新创建输入并尝试消除噪声。DAE 中的损失函数将去噪后重建的输出与原始未损坏的输入进行比较。
- **卷积自编码器（CAE）**：之前讨论的自编码器使用全连接层，这种模式类似于多层感知器模型。我们也可以使用卷积层而不是全连接层（或稠密层）。当使用卷积层创建自编码器时，它被称为 CAE。例如，我们可以为 CAE 提供以下层：

  输入 → 卷积 → 池化 → 卷积 → 池化 → 输出

  第一组卷积和池化层可以看成是编码器，它将高维的输入特征变为低维特征。第二组卷积和池化层可以看成是解码器，它将低维特征转换成高维特征。
- **变分自编码器（VAE）**：变分自编码器的架构是自编码器领域的最新研究方向。VAE 是一种生成模型，它产生概率分布的参数，从中可以生成原始数据或与原始数据非常相似的数据。

在 VAE 中，编码器将输入样本转换为潜在（latent）空间中的参数，使用该参数对潜在点采样。然后解码器使用潜点重新生成原始输入数据。因此，VAE 的重点是最大化输入数据的概率，而不是试图从输出重新构造输入。

下面会在 TensorFlow 和 Keras 中构建自编码器。我们将在 MNIST 数据集上构建自编码器。自编码器将学习用较少的神经元（或特征）表示 MNIST 数据集的手写数字。

读者可以按照 Jupyter 笔记本 ch-10_AutoEncoders_TF_and_Keras 中的代码进行操作。

像以前一样，首先使用以下代码读取 MNIST 数据集：

```
from tensorflow.examples.tutorials.mnist.input_data import input_data
dataset_home = os.path.join(datasetslib.datasets_root,'mnist')
mnist = input_data.read_data_sets(dataset_home,one_hot=False)

X_train = mnist.train.images
X_test = mnist.test.images
Y_train = mnist.train.labels
Y_test = mnist.test.labels

pixel_size = 28
```

从训练数据集和测试数据集中提取 4 个不同的图像及它们各自的标签：

```
while True:
    train_images,train_labels = mnist.train.next_batch(4)
    if len(set(train_labels))==4:
        break
while True:
    test_images,test_labels = mnist.test.next_batch(4)
    if len(set(test_labels))==4:
        break
```

下面介绍如何使用 MNIST 数据集构建自编码器。

 可以按照 Jupyter 笔记本 ch-10_AutoEncoders_TF_and_Keras 中的代码进行操作。

10.2 基于 TensorFlow 的堆叠自编码器

在 TensorFlow 中构建堆叠自编码器模型的步骤如下：

1）首先，定义超参数如下：

```
learning_rate = 0.001
n_epochs = 20
batch_size = 100
n_batches = int(mnist.train.num_examples/batch_size)
```

2）定义输入（即特征）和输出（即目标）的数量。输出数量与输入数量相同：

```
# MNIST图像中的像素数作为输入数
n_inputs = 784
n_outputs = n_inputs
```

3）定义输入图像和输出图像的占位符：

```
x = tf.placeholder(dtype=tf.float32, name="x", shape=[None, n_inputs])
y = tf.placeholder(dtype=tf.float32, name="y", shape=[None, n_outputs])
```

4）为编码器层和解码器层指定神经元数量 [512, 256, 256, 512]：

```
# 隐藏层数
n_layers = 2
# 每个隐藏层中的神经元
n_neurons = [512,256]
# 添加解码器层数
n_neurons.extend(list(reversed(n_neurons)))
n_layers = n_layers * 2
```

5）定义参数 w 和 b：

```
w=[]
b=[]

for i in range(n_layers):
    w.append(tf.Variable(tf.random_normal([n_inputs \
                if i==0 else n_neurons[i-1],n_neurons[i]]),
                    name="w_{0:04d}".format(i)
                )
            )
    b.append(tf.Variable(tf.zeros([n_neurons[i]]),
```

```
                    name="b_{0:04d}".format(i)
                )
            )
    w.append(tf.Variable(tf.random_normal([n_neurons[n_layers-1] \
                        if n_layers > 0 else n_inputs,n_outputs]),
                         name="w_out"
                )
            )
    b.append(tf.Variable(tf.zeros([n_outputs]),name="b_out"))
```

6）构建网络，每一层使用 sigmoid 作为激活函数：

```
# x是输入层
layer = x
# add hidden layers
for i in range(n_layers):
    layer = tf.nn.sigmoid(tf.matmul(layer, w[i]) + b[i])
# 添加输出层
layer = tf.nn.sigmoid(tf.matmul(layer, w[n_layers]) + b[n_layers])
model = layer
```

7）使用 AdamOptimizer 作为优化器，损失函数为 mean_squared_error：

```
mse = tf.losses.mean_squared_error
loss = mse(predictions=model, labels=y)
optimizer = tf.train.AdamOptimizer(learning_rate=learning_rate)
optimizer = optimizer.minimize(loss)
```

8）训练模型并在训练数据集和测试数据集上进行预测：

```
with tf.Session() as tfs:
    tf.global_variables_initializer().run()
    for epoch in range(n_epochs):
        epoch_loss = 0.0
        for batch in range(n_batches):
            X_batch, _ = mnist.train.next_batch(batch_size)
            feed_dict={x: X_batch,y: X_batch}
            _,batch_loss = tfs.run([optimizer,loss], feed_dict)
            epoch_loss += batch_loss
        if (epoch%10==9) or (epoch==0):
            average_loss = epoch_loss / n_batches
            print('epoch: {0:04d} loss = {1:0.6f}'
                  .format(epoch,average_loss))
    # 使用训练的自编码器模型预测图像
    Y_train_pred = tfs.run(model, feed_dict={x: train_images})
    Y_test_pred = tfs.run(model, feed_dict={x: test_images})
```

9）在 20 次迭代之后，损失函数明显减少，会得到如下输出结果：

```
epoch: 0000    loss = 0.156696
epoch: 0009    loss = 0.091367
epoch: 0019    loss = 0.078550
```

10）现在已经训练好模型，下面通过训练好的模型得到生成图像，并显示这些图像。display_images 是一个辅助函数，它用来显示图像：

```
import random

# 用于显示图像和标签的函数
# 图像应采用NHW或NHWC格式
def display_images(images, labels, count=0, one_hot=False):
    # 如果未提供要显示的图像数量，则显示所有图像
    if (count==0):
        count = images.shape[0]

    idx_list = random.sample(range(len(labels)),count)
    for i in range(count):
        plt.subplot(4, 4, i+1)
        plt.title(labels[i])
        plt.imshow(images[i])
        plt.axis('off')
    plt.tight_layout()
    plt.show()
```

使用这个函数显示训练集中的 4 个图像和自编码器生成的图像。第一行为真实的图像，第二行为生成的图像，如图 10-1 所示。

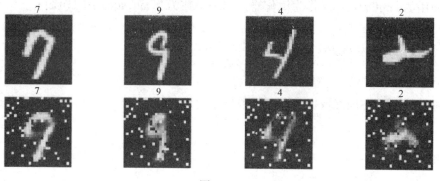

图 10-1

生成的图像有一点点噪声，可以通过更多的训练和调整超参数来去除这些噪声。生成的图像还不错，因为是在用训练集上的图像来训练自编码器，所以它们之间有一定关系。下面是预测测试集图像的结果。第一行为实际的图像，第二行为生成的图像，如果 10-2 所示。

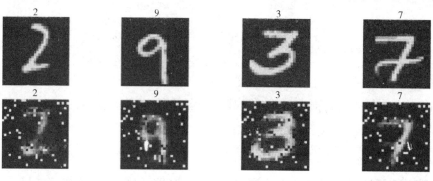

图 10-2

经过训练的自编码器能够用 256 个特征（从 768 个特征中学到的）生成相同的数字。可以通过调整超参数和采用更多的训练次数消除生成图像中的噪声。

10.3 基于 Keras 的堆叠自编码器

现在用 Keras 构建相同的自编码器。

我们使用以下命令清除 Jupyter 笔记本中的计算图，从而可以构建一个新的计算图，该计算图不会覆盖上一个会话或计算图中的任何内存：
```
tf.reset_default_graph()
keras.backend.clear_session()
```

1）首先，导入 keras 库并定义超参数和网络层：

```
import keras
from keras.layers import Dense
from keras.models import Sequential

learning_rate = 0.001
n_epochs = 20
batch_size = 100
n_batches = int(mnist.train.num_examples/batch_sizee
# MNIST图像中的像素数量作为输入数
n_inputs = 784
n_outputs = n_i
# 隐藏层数
n_layers = 2
# 每个隐藏层中的神经元
n_neurons = [512,256]
# 添加解码器层数
n_neurons.extend(list(reversed(n_neurons)))
n_layers = n_layers * 2
```

2）接下来，构建一个顺序模型并为其添加一个全连接层。有一点变化：对隐藏层使用 relu 激活函数，对最终层使用线性激活函数：

```
model = Sequential()

# 将输入添加到第一层
model.add(Dense(units=n_neurons[0], activation='relu',
    input_shape=(n_inputs,)))

for i in range(1,n_layers):
    model.add(Dense(units=n_neurons[i], activation='relu'))

# 将最终层添加为输出层
model.add(Dense(units=n_outputs, activation='linear'))
```

3）下面是模型的摘要：

```
model.summary()
```

该模型有 5 个全连接层，共有 1 132 816 个参数：

```
Layer (type)                 Output Shape              Param #
=================================================================
dense_1 (Dense)              (None, 512)               401920
_____
dense_2 (Dense)              (None, 256)               131328
_____
dense_3 (Dense)              (None, 256)               65792
_____
dense_4 (Dense)              (None, 512)               131584
_____
dense_5 (Dense)              (None, 784)               402192
=================================================================
Total params: 1,132,816
Trainable params: 1,132,816
Non-trainable params: 0
_____
```

4）指定均方误差函数为损失函数，然后编译模型：

```
model.compile(loss='mse',
    optimizer=keras.optimizers.Adam(lr=learning_rate),
    metrics=['accuracy'])

model.fit(X_train, X_train,batch_size=batch_size,
    epochs=n_epochs)
```

在 20 次迭代之后，损失函数值为 0.0046，而之前的损失函数值为 0.078550：

```
Epoch 1/20
55000/55000 [==========================] - 18s - loss: 0.0193 - acc: 0.0117
Epoch 2/20
55000/55000 [==========================] - 18s - loss: 0.0087 - acc: 0.0139
...
...
...
Epoch 20/20
55000/55000 [==========================] - 16s - loss: 0.0046 - acc: 0.0171
```

现在让我们预测并显示模型生成的训练图像和测试图像。第一行为真实的图像，第二行为生成的图像。图 10-3 为基于训练集生成的图像。

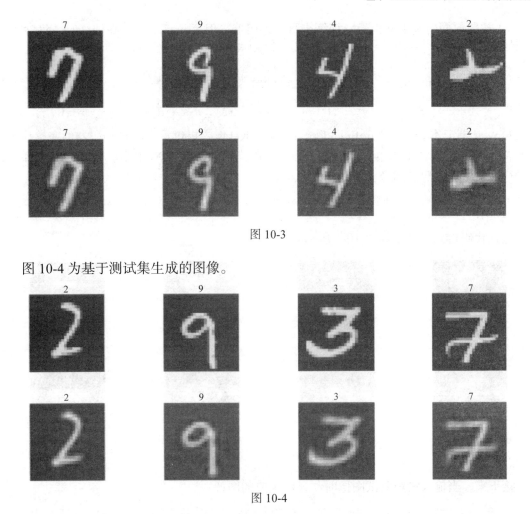

图 10-3

图 10-4 为基于测试集生成的图像。

图 10-4

能够从 256 个特征生成这些如此相似的图像已经非常不错了。

10.4 基于 TensorFlow 的去噪自编码器

正如本章 10.1 节介绍的，可以使用去噪自编码器训练模型，即通过去除输入图像中的噪声训练模型：

1）编写辅助函数为图像添加噪声：

```
def add_noise(X):
    return X + 0.5 * np.random.randn(X.shape[0],X.shape[1])
```

2）在测试图像中添加噪声，并将其存储在一个单独的列表中：

```
test_images_noisy = add_noise(test_images)
```

这些测试图像将用来测试去噪模型。

3）按照前面的例子来构建、训练去噪自编码器，但有一点不同：在训练时，将含噪声的图像作为输入，并用不含噪声的图像检查重建误差和去噪误差，如下面的代码所示：

```
X_batch, _ = mnist.train.next_batch(batch_size)
X_batch_noisy = add_noise(X_batch)
feed_dict={x: X_batch_noisy, y: X_batch}
_,batch_loss = tfs.run([optimizer,loss], feed_dict=feed_dict)
```

Jupyter 笔记本 ch-10_AutoEncoders_TF_and_Keras 中提供了去噪自编码器的完整代码。首先显示从 DAE 模型生成的测试图像；下面代码的第一行表示原始无噪声的测试图像，第二行表示生成的测试图像：

```
display_images(test_images.reshape(-1,pixel_size,pixel_size),test_labels)
display_images(Y_test_pred1.reshape(-1,pixel_size,pixel_size),test_labels)
```

上述代码得到的结果如图 10-5 所示。

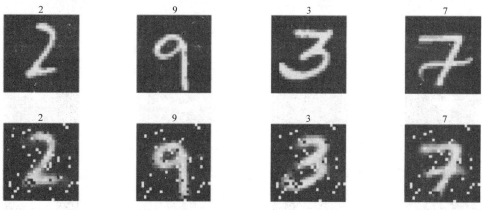

图 10-5

接下来，当输入噪声测试图像时，显示生成的图像：

```
display_images(test_images_noisy.reshape(-1,pixel_size,pixel_size),
    test_labels)
display_images(Y_test_pred2.reshape(-1,pixel_size,pixel_size),test_labels)
```

上述代码的结果如图 10-6 所示。

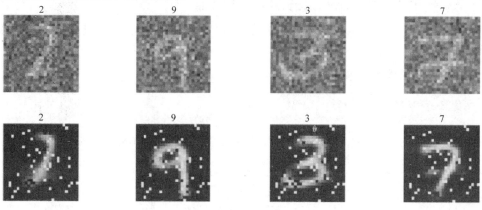

图 10-6

这太酷了！即使该图像包含大量噪声，上述模型也可以学习图像并生成几乎完全正确的图像。通过适当调整超参数可以进一步提高生成图像的质量。

10.5 基于 Keras 的去噪自编码器

现在在 Keras 中构建同样的去噪自编码器。

由于 Keras 负责按批量大小输入训练数据集，因此，在模型中创建了一个包含噪声的训练集作为输入：

```
X_train_noisy = add_noise(X_train)
```

在 Jupyter 笔记本 ch-10_AutoEncoders_TF_and_Keras 中提供了 Keras DAE 的完整代码。基于 Keras 的 DAE 模型摘要如下：

```
Layer (type)                 Output Shape              Param #
=================================================================
dense_1 (Dense)              (None, 512)               401920
_____
dense_2 (Dense)              (None, 256)               131328
_____
dense_3 (Dense)              (None, 256)               65792
_____
dense_4 (Dense)              (None, 512)               131584
_____
dense_5 (Dense)              (None, 784)               402192
=================================================================
Total params: 1,132,816
Trainable params: 1,132,816
Non-trainable params: 0
```

由于 DAE 模型很复杂，必须将迭代周期增加到 100 才能很好地训练模型：

```
n_epochs=100

model.fit(x=X_train_noisy, y=X_train,
    batch_size=batch_size,
    epochs=n_epochs,
    verbose=0)

Y_test_pred1 = model.predict(test_images)
Y_test_pred2 = model.predict(test_images_noisy)
```

输出生成的图像：

```
display_images(test_images.reshape(-1,pixel_size,pixel_size),test_labels)
display_images(Y_test_pred1.reshape(-1,pixel_size,pixel_size),test_labels)
```

第一行是原始的测试图像，第二行是生成的测试图像，如图 10-7 所示。

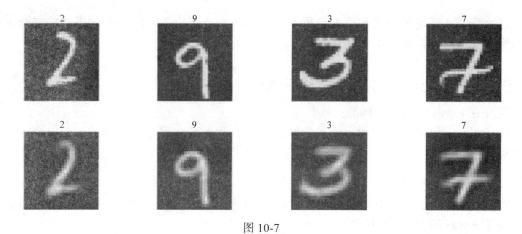

图 10-7

```
display_images(test_images_noisy.reshape(-1,pixel_size,pixel_size),
    test_labels)
display_images(Y_test_pred2.reshape(-1,pixel_size,pixel_size),test_labels)
```

第一行是含噪声的测试图像，第二行是生成的测试图像，如图 10-8 所示。

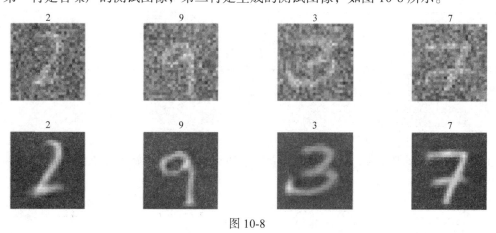

图 10-8

正如所看到的，去噪自编码器可以很好地从具有噪声的图像中生成图像。

10.6 基于 TensorFlow 的变分自编码器

变分自编码器是自编码器的现代生成（generative）版本，接下来为前面的问题构建一个变分自编码器。将通过原始测试集和包含噪声测试集的图像测试变分自编码器。

这里将使用不同的编码风格构建此自编码器，这只是为了展示 TensorFlow 的不同编码风格：

1）首先，定义超参数：

```
learning_rate = 0.001
n_epochs = 20
batch_size = 100
```

```
n_batches = int(mnist.train.num_examples/batch_size)
# MNIST图像中的像素数量作为输入数
n_inputs = 784
n_outputs = n_inputs
```

2）接下来，定义参数字典保存权重和偏差参数：

```
params={}
```

3）定义每个编码器和解码器中隐藏层的数量：

```
n_layers = 2
# 每个隐藏层中的神经元
n_neurons = [512,256]
```

4）变分编码器新增加了一个潜在变量 z 的维数：

```
n_neurons_z = 128  # 潜在变量的维数
```

5）使用 tanh 作为激活函数：

```
activation = tf.nn.tanh
```

6）定义输入和输出占位符：

```
x = tf.placeholder(dtype=tf.float32, name="x",
                   shape=[None, n_inputs])
y = tf.placeholder(dtype=tf.float32, name="y",
                   shape=[None, n_outputs])
```

7）定义输入层：

```
# x是输入层
layer = x
```

8）定义编码器网络的偏差和权重，并创建网络层。变分自编码器中的编码器网络也称为识别网络（或推理网络或概率编码器网络）：

```
for i in range(0,n_layers):
    name="w_e_{0:04d}".format(i)
    params[name] = tf.get_variable(name=name,
        shape=[n_inputs if i==0 else n_neurons[i-1],
        n_neurons[i]],
        initializer=tf.glorot_uniform_initializer()
        )
    name="b_e_{0:04d}".format(i)
    params[name] = tf.Variable(tf.zeros([n_neurons[i]]),
        name=name
        )
    layer = activation(tf.matmul(layer,
        params["w_e_{0:04d}".format(i)]
        ) + params["b_e_{0:04d}".format(i)]
        )
```

9）为潜在变量的均值和方差添加一个层：

```
name="w_e_z_mean"
params[name] = tf.get_variable(name=name,
```

```
            shape=[n_neurons[n_layers-1], n_neurons_z],
            initializer=tf.glorot_uniform_initializer()
            )
name="b_e_z_mean"
params[name] = tf.Variable(tf.zeros([n_neurons_z]),
    name=name
    )
z_mean = tf.matmul(layer, params["w_e_z_mean"]) +
            params["b_e_z_mean"]
name="w_e_z_log_var"
params[name] = tf.get_variable(name=name,
    shape=[n_neurons[n_layers-1], n_neurons_z],
    initializer=tf.glorot_uniform_initializer()
    )
name="b_e_z_log_var"
params[name] = tf.Variable(tf.zeros([n_neurons_z]),
    name="b_e_z_log_var"
    )
z_log_var = tf.matmul(layer, params["w_e_z_log_var"]) +
            params["b_e_z_log_var"]
```

10）epsilon 表示与 z 方差变量保持相同形状的噪声分布：

```
epsilon = tf.random_normal(tf.shape(z_log_var),
    mean=0,
    stddev=1.0,
    dtype=tf.float32,
    name='epsilon'
    )
```

11）根据均值、对数方差和噪声定义后验分布：

```
z = z_mean + tf.exp(z_log_var * 0.5) * epsilon
```

12）定义解码器网络的权重和偏差，并添加解码器层。变分自编码器中的解码器网络也称为概率解码器或生成器（generator）网络：

```
#  添加生成器/概率解码器网络参数和层
layer = z

for i in range(n_layers-1,-1,-1):
name="w_d_{0:04d}".format(i)
    params[name] = tf.get_variable(name=name,
        shape=[n_neurons_z if i==n_layers-1 else n_neurons[i+1],
        n_neurons[i]],
        initializer=tf.glorot_uniform_initializer()
        )
name="b_d_{0:04d}".format(i)
params[name] = tf.Variable(tf.zeros([n_neurons[i]]),
    name=name
    )
layer = activation(tf.matmul(layer,
    params["w_d_{0:04d}".format(i)]) +
    params["b_d_{0:04d}".format(i)])
```

13）最后，定义输出层：

```
name="w_d_z_mean"
params[name] = tf.get_variable(name=name,
    shape=[n_neurons[0],n_outputs],
    initializer=tf.glorot_uniform_initializer()
    )
name="b_d_z_mean"
    params[name] = tf.Variable(tf.zeros([n_outputs]),
    name=name
    )
name="w_d_z_log_var"
params[name] = tf.Variable(tf.random_normal([n_neurons[0],
    n_outputs]),
    name=name
    )
name="b_d_z_log_var"
    params[name] = tf.Variable(tf.zeros([n_outputs]),
    name=name
    )
layer = tf.nn.sigmoid(tf.matmul(layer, params["w_d_z_mean"]) +
    params["b_d_z_mean"])

model = layer
```

14）变分自编码器的损失函数由重构损失函数和正则化损失函数组成。因此，将损失函数定义为重构损失函数和正则化损失函数之和：

```
rec_loss = -tf.reduce_sum(y * tf.log(1e-10 + model) + (1-y)
                    * tf.log(1e-10 + 1 - model), 1)
reg_loss = -0.5*tf.reduce_sum(1 + z_log_var - tf.square(z_mean)
                    - tf.exp(z_log_var), 1)
loss = tf.reduce_mean(rec_loss+reg_loss)
```

15）指定优化器函数为 AdapOptimizer：

```
optimizer = tf.train.AdamOptimizer(learning_rate=learning_rate)
        .minimize(loss)
```

16）现在训练模型并通过非噪声和噪声测试图像来生成图像：

```
with tf.Session() as tfs:
    tf.global_variables_initializer().run()
    for epoch in range(n_epochs):
        epoch_loss = 0.0
        for batch in range(n_batches):
            X_batch, _ = mnist.train.next_batch(batch_size)
            feed_dict={x: X_batch,y: X_batch}
            _,batch_loss = tfs.run([optimizer,loss],
                    feed_dict=feed_dict)
            epoch_loss += batch_loss
        if (epoch%10==9) or (epoch==0):
            average_loss = epoch_loss / n_batches
            print("epoch: {0:04d} loss = {1:0.6f}"
```

```
                .format(epoch,average_loss))
# 使用自编码器模型训练预测图像
Y_test_pred1 = tfs.run(model, feed_dict={x: test_images})
Y_test_pred2 = tfs.run(model, feed_dict={x: test_images_noisy})
```

得到如下结果：

```
epoch: 0000    loss = 180.444682
epoch: 0009    loss = 106.817749
epoch: 0019    loss = 102.580904
```

现在展示图像：

```
display_images(test_images.reshape(-1,pixel_size,pixel_size),test_labels)
display_images(Y_test_pred1.reshape(-1,pixel_size,pixel_size),test_labels)
```

结果如图 10-9 所示。

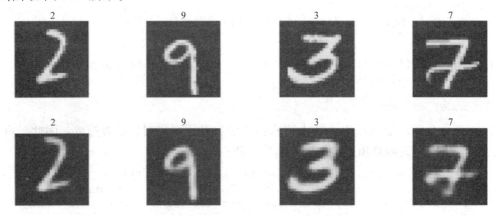

图 10-9

```
display_images(test_images_noisy.reshape(-1,pixel_size,pixel_size),
    test_labels)
display_images(Y_test_pred2.reshape(-1,pixel_size,pixel_size),test_labels)
```

结果如图 10-10 所示。

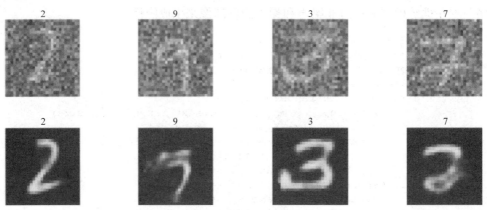

图 10-10

同样可通过超参数调整和增加迭代次数来改善结果。

10.7 基于 Keras 的变分自编码器

在 Keras 中,构建变分自编码器更容易,所用的代码量更少。基于 Keras 的变分自编码器最好使用功能方式构建。到目前为止,本章使用 Keras 的顺序方式构建模型。这个例子将用 Keras 的功能方式构建变分自编码器。在 Keras 中构建变分自编码器的步骤如下:

1)定义隐藏层和潜在变量层中的超参数和神经元数量:

```
import keras
from keras.layers import Lambda, Dense, Input, Layer
from keras.models import Model
from keras import backend as K

learning_rate = 0.001
batch_size = 100
n_batches = int(mnist.train.num_examples/batch_size)
# MNIST图像中的像素数量作为输入数
n_inputs = 784
n_outputs = n_inputs
# 隐藏层数
n_layers = 2
# 每个隐藏层中的神经元
n_neurons = [512,256]
# 潜在变量的维度
n_neurons_z = 128
```

2)构建输入层:

```
x = Input(shape=(n_inputs,), name='input')
```

3)构建编码器层,以及潜在变量的均值和方差层:

```
# 构建编码器
layer = x
for i in range(n_layers):
    layer = Dense(units=n_neurons[i],
activation='relu',name='enc_{0}'.format(i))(layer)

z_mean = Dense(units=n_neurons_z,name='z_mean')(layer)
z_log_var = Dense(units=n_neurons_z,name='z_log_v')(layer)
```

4)创建噪声和后验分布:

```
# 噪声分布
epsilon = K.random_normal(shape=K.shape(z_log_var),
        mean=0,stddev=1.0)

# 后验分布
z = Lambda(lambda zargs: zargs[0] + K.exp(zargs[1] * 0.5) *
epsilon,
    name='z')([z_mean,z_log_var])
```

5)添加解码器层：

```
# 添加生成器/概率解码器网络层
layer = z
for i in range(n_layers-1,-1,-1):
    layer = Dense(units=n_neurons[i], activation='relu',
            name='dec_{0}'.format(i))(layer)
```

6)定义最终输出层：

```
y_hat = Dense(units=n_outputs, activation='sigmoid',
        name='output')(layer)
```

7)最后，从输入层和输出层定义模型并显示模型摘要：

```
model = Model(x,y_hat)
model.summary()
```

下面是模型摘要：

Layer (type)	Output Shape	Param #	Connected to
input (InputLayer)	(None, 784)	0	
enc_0 (Dense)	(None, 512)	401920	input[0][0]
enc_1 (Dense)	(None, 256)	131328	enc_0[0][0]
z_mean (Dense)	(None, 128)	32896	enc_1[0][0]
z_log_v (Dense)	(None, 128)	32896	enc_1[0][0]
z (Lambda)	(None, 128)	0	z_mean[0][0] z_log_v[0][0]
dec_1 (Dense)	(None, 256)	33024	z[0][0]
dec_0 (Dense)	(None, 512)	131584	dec_1[0][0]
output (Dense)	(None, 784)	402192	dec_0[0][0]

```
Total params: 1,165,840
Trainable params: 1,165,840
Non-trainable params: 0
```

8)定义一个由重构函数与正则化损失函数相加而成的损失函数：

```
def vae_loss(y, y_hat):
    rec_loss = -K.sum(y * K.log(1e-10 + y_hat) + (1-y) *
            K.log(1e-10 + 1 - y_hat), axis=-1)
    reg_loss = -0.5 * K.sum(1 + z_log_var - K.square(z_mean) -
            K.exp(z_log_var), axis=-1)
    loss = K.mean(rec_loss+req_loss)
    return loss
```

9）使用此损失函数编译模型：

```
model.compile(loss=vae_loss,
    optimizer=keras.optimizers.Adam(lr=learning_rate))
```

10）训练50个迭代周期的模型，并预测图像，正如在前面部分所做的那样：

```
n_epochs=50
model.fit(x=X_train_noisy,y=X_train,batch_size=batch_size,
    epochs=n_epochs,verbose=0)
Y_test_pred1 = model.predict(test_images)
Y_test_pred2 = model.predict(test_images_noisy)
```

显示图像结果：

```
display_images(test_images.reshape(-1,pixel_size,pixel_size),test_labels)
display_images(Y_test_pred1.reshape(-1,pixel_size,pixel_size),test_labels)
```

结果如图10-11所示。

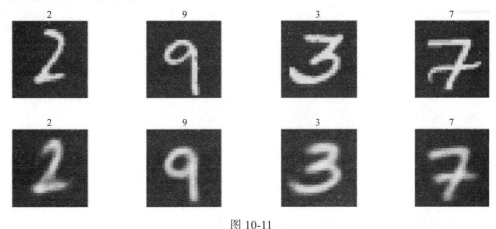

图10-11

```
display_images(test_images_noisy.reshape(-1,pixel_size,pixel_size),
    test_labels)
display_images(Y_test_pred2.reshape(-1,pixel_size,pixel_size),test_labels)
```

结果如图10-12所示。

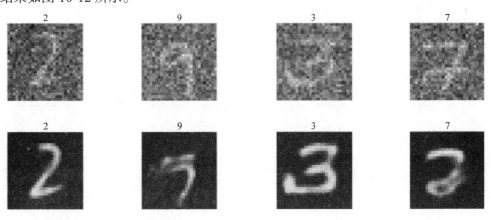

图10-12

这个结果很棒！！生成的图像更清晰了。

10.8 总结

  自编码器是无监督数据学习的绝佳工具，通常用于降低数据的维数，因此数据可以用较少数量的特征表示。本章介绍了各种类型的自编码器。使用 TensorFlow 和 Keras 构建三种类型的自编码器：堆叠自编码器、去噪自编码器和变分自编码器，这些都是基于 MNIST 数据集完成的。

  最近这几章介绍了如何使用 TensorFlow 和 Keras 构建各种机器学习和深度学习模型，例如回归、分类、MLP、CNN、RNN 和自编码器。下一章将介绍 TensorFlow 和 Keras 的高级功能，这些功能允许人们将模型应用于生产环境中。

第 11 章 使用 TF 服务提供生成环境下的 TensorFlow 模型

在开发环境训练和验证了 TensorFlow 模型之后,就需要发布模型。在发布模型之后,TensorFlow 模型需要托管(host)在某个地方,供应用工程师和软件工程师集成到各种应用程序中。因此,TensorFlow 提供了一个高性能服务器,即 TensorFlow 服务。

要为生产环境中提供 TensorFlow 模型,需要离线训练并保存模型,然后在生产环境中恢复经过训练的模型。保存的 TensorFlow 模型由以下文件组成:

- **元计算图** (meta-graph):计算图的协议缓冲区定义,元计算图保存在扩展名为 .meta 的文件中。
- **检查点** (checkpoint):各种变量的值。检查点保存在两个文件中:一个文件的扩展名为 .index,另一个文件的扩展名为 .data-00000-of-00001。

本章将学习保存和恢复模型的各种方法,也会介绍如何使用 TF 服务提供(serve)模型,为了简化操作,这里将使用 MNIST 示例。本章将介绍以下主题:

- 在 TensorFlow 中保存和恢复模型;
- 保存和恢复 Keras 模型;
- TensorFlow 服务;
- 安装 TF 服务;
- 保存 TF 服务的模型;
- 使用 TF 服务提供服务模型;
- 在 Docker 容器中提供 TF 服务;
- 基于 Kubernetes 的 TF 服务。

11.1 在 TensorFlow 中保存和恢复模型

在 TensorFlow 中,可以通过以下两种方法保存和恢复模型以及相关的变量:

- 通过 tf.train.Saver 类创建的 saver 对象;
- 通过 tf.saved_model_builder.SavedModelBuilder 类创建 SavedModel 格式的对象。

下面介绍这两种方法的实际应用。

 可以按照 Jupyter 笔记本 ch-11a_Saving_and_Restoring_TF_Models 中的代码进行操作。

11.1.1 使用 saver 类保存和恢复所有网络计算图变量

操作如下：

1）要使用 saver 类，首先要创建该类的对象实例：

```
saver = tf.train.Saver()
```

2）保存计算图中所有变量的最简单方法是调用 save ( ) 方法，该方法有 2 个参数：session 对象和保存变量的文件路径：

```
with tf.Session() as tfs:
    ...
    saver.save(tfs,"saved-models/model.ckpt")
```

3）要恢复变量，调用 restore ( ) 方法：

```
with tf.Session() as tfs:
    saver.restore(tfs,"saved-models/model.ckpt")
    ...
```

4）修改本书第 1 章的示例，下面的代码将保存该简单示例中计算图变量：

```
# 假设线性模型为 y = w * x + b
# 定义模型参数
w = tf.Variable([.3], tf.float32)
b = tf.Variable([-.3], tf.float32)
# Define model input and output
x = tf.placeholder(tf.float32)
y = w * x + b
output = 0

# 创建saver对象
saver = tf.train.Saver()

with tf.Session() as tfs:
    # 初始化和输出变量y
    tfs.run(tf.global_variables_initializer())
    output = tfs.run(y,{x:[1,2,3,4]})
    saved_model_file = saver.save(tfs,
        'saved-models/full-graph-save-example.ckpt')
    print('Model saved in {}'.format(saved_model_file))
    print('Values of variables w,b: {}{}'
        .format(w.eval(),b.eval()))
    print('output={}'.format(output))
```

可以得到以下输出：

```
Model saved in saved-models/full-graph-save-example.ckpt
Values of variables w,b: [ 0.30000001][-0.30000001]
output=[ 0.          0.30000001  0.60000002  0.90000004]
```

5）从刚刚创建的检查点文件中恢复变量：

```
# 假设线性模型为 y = w * x + b
# 定义模型参数
w = tf.Variable([0], dtype=tf.float32)
b = tf.Variable([0], dtype=tf.float32)
# 定义模型输入和输出
x = tf.placeholder(dtype=tf.float32)
y = w * x + b
output = 0

# 创建saver对象
saver = tf.train.Saver()

with tf.Session() as tfs:
    saved_model_file = saver.restore(tfs,
        'saved-models/full-graph-save-example.ckpt')
    print('Values of variables w,b: {}{}'
        .format(w.eval(),b.eval()))
    output = tfs.run(y,{x:[1,2,3,4]})
    print('output={}'.format(output))
```

注意，在恢复代码时并没有调用 tf.global_variables_initializer( )，因为变量会从文件中恢复，所以没有必要初始化变量。这些代码会得到以下输出结果，这个结果是根据恢复的变量计算得出的：

```
INFO:tensorflow:Restoring parameters from saved-models/full-graph-save-
example.ckpt
Values of variables w,b: [ 0.30000001][-0.30000001]
output=[ 0.          0.30000001  0.60000002  0.90000004]
```

11.1.2 使用 saver 类保存和恢复所选变量

默认情况下，Saver( ) 类将保存神经网络中的所有变量。为了保存指定变量，可在实例化 Saver() 类时，将变量列表传递给 Saver( ) 类的构造函数：

```
# 创建saver对象
saver = tf.train.Saver({'weights': w})
```

变量名可通过列表或字典进行传递，如果变量名通过列表传递，则列表中的每个变量将以自己的名称保存。变量也可以通过由键值对组成的字典进行传递，其中键表示要保存的名称，值表示要保存的变量名。

下面的代码与上面的代码差不多，所不同的是这里只保存权重变量 w，在保存时将其命名为 weights：

```
# 在TensorFlow中保存计算图中的选定变量

# 假设线性模型y = w * x + b
# 定义模型参数
w = tf.Variable([.3], tf.float32)
b = tf.Variable([-.3], tf.float32)
# 定义模型输入和输出
```

```python
x = tf.placeholder(tf.float32)
y = w * x + b
output = 0

# 创建saver对象
saver = tf.train.Saver({'weights': w})

with tf.Session() as tfs:
    # 初始化和输出变量y
    tfs.run(tf.global_variables_initializer())
    output = tfs.run(y,{x:[1,2,3,4]})
    saved_model_file = saver.save(tfs,
        'saved-models/weights-save-example.ckpt')
    print('Model saved in {}'.format(saved_model_file))
    print('Values of variables w,b: {}{}'
        .format(w.eval(),b.eval()))
    print('output={}'.format(output))
```

可以得到以下输出结果:

```
Model saved in saved-models/weights-save-example.ckpt
Values of variables w,b: [ 0.30000001][-0.30000001]
output=[ 0.          0.30000001  0.60000002  0.90000004]
```

检查点文件只保存权重变量,没有保存偏差。现在将偏差和权重初始化为零,并恢复权重,具体代码如下:

```python
# 在TensorFlow中恢复计算图中所选定变量
tf.reset_default_graph()
# 假设线性模型为y = w * x + b
# 定义模型参数
w = tf.Variable([0], dtype=tf.float32)
b = tf.Variable([0], dtype=tf.float32)
# 定义模型输入和输出
x = tf.placeholder(dtype=tf.float32)
y = w * x + b
output = 0

# 创建saver对象
saver = tf.train.Saver({'weights': w})

with tf.Session() as tfs:
    b.initializer.run()
    saved_model_file = saver.restore(tfs,
        'saved-models/weights-save-example.ckpt')
    print('Values of variables w,b: {}{}'
        .format(w.eval(),b.eval()))
    output = tfs.run(y,{x:[1,2,3,4]})
    print('output={}'.format(output))
```

这里必须使用b.initializer.run( )初始化偏差。不使用tfs.run (tf.global_variables_initializer( ))的原因是: tfs.run会初始化所有变量,但这里并不需要初始化权重,因为权重将从检查点文件中恢复。

可以得到以下输出结果,此次计算仅使用恢复的权重,而偏差被设置为零:

```
INFO:tensorflow:Restoring parameters from saved-models/weights-save-
example.ckpt
Values of variables w,b: [ 0.30000001][ 0.]
output=[ 0.30000001  0.60000002  0.90000004  1.20000005]
```

11.2 保存和恢复 Keras 模型

在 Keras 中，保存和恢复模型非常简单，Keras 提供三种选择：
- 保存整个模型，包括网络体系结构、权重（参数）、训练配置和优化器状态；
- 只保存网络结构；
- 只保存权重。

要保存完整模型，可使用 model.save (filepath) 函数，该函数将完整的模型保存在 HDF5 文件中。保存的模型可使用 keras.models.load_model (filepath) 函数加载。该函数可以加载所有内容并能编译模型。

要保存模型的结构，请使用 model.to_json ( ) 或 model.to_yaml ( ) 函数，这些函数会返回一个可以写入磁盘文件的字符串。在恢复模型结构时，可以使用 keras.models.model_from_json (json_string) 或 keras.models.model_from_yaml (yaml_string) 函数读取字符串并恢复模型结构。这两个函数可以返回一个模型实例。

要保存模型的权重，请使用 model.save_weights (path_to_h5_file) 函数。权重可以通过 model.load_weights (path_to_h5_file) 函数恢复。

11.3 TensorFlow 服务

TensorFlow 服务（TFS）是一种高性能服务器架构，用于为生产中的机器学习模型提供服务，TFS 能方便地集成 TensorFlow 所构建的模型。

在 TFS 中，模型由一个或多个 servables 组成。这些 servables 用于执行计算，例如：
- 针对嵌入查找的查找表；
- 返回预测结果的单个模型；
- 返回预测元组的元组模型；
- 查找表或模型的分片（shard）。

manager 组件管理 servables 的整个生命周期，包括加载 / 卸载 servables 以及为 servables 提供服务。

 以下链接介绍了 TensorFlow 服务的内部架构和工作流程：https://www.tensorflow.org/serving/architecture_overview。

11.3.1 安装 TF 服务

可在 Ubuntu 上使用 aptitude 安装 TensorFlow 的 ModelServer。
1）首先，在 shell 提示符下使用以下命令添加 TensorFlow 服务发布的 URI 作为包源（一

次性安装）:

```
$ echo "deb [arch=amd64] http://storage.googleapis.com/tensorflow-serving-apt stable tensorflow-model-server tensorflow-model-server-universal" | sudo tee /etc/apt/sources.list.d/tensorflow-serving.list

$ curl https://storage.googleapis.com/tensorflow-serving-apt/tensorflow-serving.release.pub.gpg | sudo apt-key add -
```

2）在 shell 提示符下使用以下命令安装和更新 TensorFlow 的 ModelServer：

```
$ sudo apt-get update && sudo apt-get install tensorflow-model-server
```

这将使用特定平台下编译器的优化指令（例如使用 SSE4 和 AVX 指令）安装 ModelServer 版本。但是，如果优化版本安装在旧计算机上不能运行，则可以安装通用版本：

```
$ sudo apt-get remove tensorflow-model-server
$ sudo apt-get update && sudo apt-get install tensorflow-model-server-universal
```

对于其他操作系统和针对源代码的安装，请参阅以下链接：https://www.tensorflow.org/serving/setup。

当有新版本的 ModelServer 发布时，可以使用以下命令升级到新版本：

```
$ sudo apt-get update && sudo apt-get upgrade tensorflow-model-server
```

3）现在已经安装了 ModelServer，可使用以下命令运行服务：

```
$ tensorflow-model-server
```

4）要连接到 tensorflow-model-server，请使用 pip 安装 python 的客户端软件包：

```
$ sudo pip2 install tensorflow-serving-api
```

TF Serving API 仅适用于 Python 2，并不适用于 Python 3。

11.3.2 保存 TF 服务的模型

为了提供模型服务，需要先保存这些模型。本节所演示的 MNIST 示例与 TensorFlow 文档中的示例略有不同，TensorFlow 文档中的示例可从以下链接获得：https://www.tensorflow.org/serving/serving_basic。

TensorFlow 团队建议使用 SavedModel 保存和恢复在 TensorFlow 中构建和训练的模型。

具体内容如下：

SavedModel 是一种中性语言的、可恢复的、密集的序列化格式。SavedModel 支持更高级别的系统和工具生成、使用和转换 TensorFlow 模型。

 可以按照 Jupyter 笔记本中的代码（参见文件 ch-11b_Saving_TF_Models_with_SavedModel_for_TF_Serving）进行操作。

按照以下方式保存模型：
1）定义模型变量：

```
model_name = 'mnist'
model_version = '1'
model_dir = os.path.join(models_root,model_name,model_version)
```

2）获取 MNIST 数据，这与第 4 章（MLP 模型）中所做的一样：

```
from tensorflow.examples.tutorials.mnist import input_data
dataset_home = os.path.join('.','mnist')
mnist = input_data.read_data_sets(dataset_home, one_hot=True)
x_train = mnist.train.images
x_test = mnist.test.images
y_train = mnist.train.labels
y_test = mnist.test.labels
pixel_size = 28
num_outputs = 10  # 0-9 digits
num_inputs = 784  # total pixels
```

3）定义 MLP 函数并返回模型：

```
def mlp(x, num_inputs, num_outputs,num_layers,num_neurons):
    w=[]
    b=[]
    for i in range(num_layers):
        w.append(tf.Variable(tf.random_normal(
            [num_inputs if i==0 else num_neurons[i-1],
            num_neurons[i]]),name="w_{0:04d}".format(i)
            )
        )
        b.append(tf.Variable(tf.random_normal(
            [num_neurons[i]]),
            name="b_{0:04d}".format(i)
            )
        )
    w.append(tf.Variable(tf.random_normal(
        [num_neurons[num_layers-1] if num_layers > 0 \
        else num_inputs, num_outputs]),name="w_out"))
    b.append(tf.Variable(tf.random_normal([num_outputs]),
        name="b_out"))

    # x是输入层
    layer = x
```

```python
# 添加隐藏层
for i in range(num_layers):
    layer = tf.nn.relu(tf.matmul(layer, w[i]) + b[i])
# 添加输出层
layer = tf.matmul(layer, w[num_layers]) + b[num_layers]
model = layer
probs = tf.nn.softmax(model)

return model,probs
```

这里的 mlp() 函数将返回模型和概率,概率是由用于模型的 softmax 激活函数得到。

4)为图像输入和目标输出定义占位符 x_p 和 y_p:

```python
# 输入图像
serialized_tf_example = tf.placeholder(tf.string,
        name='tf_example')
feature_configs = {'x': tf.FixedLenFeature(shape=[784],
        dtype=tf.float32),}
tf_example = tf.parse_example(serialized_tf_example,
        feature_configs)
# 使用 tf.identity() 指定名称
x_p = tf.identity(tf_example['x'], name='x_p')
# 目标输出
y_p = tf.placeholder(dtype=tf.float32, name="y_p",
        shape=[None, num_outputs])
```

5)创建模型、损失函数、优化器、精度和训练函数:

```python
num_layers = 2
num_neurons = []
for i in range(num_layers):
    num_neurons.append(256)

learning_rate = 0.01
n_epochs = 50
batch_size = 100
n_batches = mnist.train.num_examples//batch_size

model,probs = mlp(x=x_p,
    num_inputs=num_inputs,
    num_outputs=num_outputs,
    num_layers=num_layers,
    num_neurons=num_neurons)

loss_op = tf.nn.softmax_cross_entropy_with_logits
loss = tf.reduce_mean(loss_op(logits=model, labels=y_p))
optimizer = tf.train.GradientDescentOptimizer(learning_rate)
train_op = optimizer.minimize(loss)

pred_check = tf.equal(tf.argmax(probs,1), tf.argmax(y_p,1))
accuracy_op = tf.reduce_mean(tf.cast(pred_check, tf.float32))

values, indices = tf.nn.top_k(probs, 10)
table = tf.contrib.lookup.index_to_string_table_from_tensor(
```

```
            tf.constant([str(i) for i in range(10)]))
prediction_classes = table.lookup(tf.to_int64(indices))
```

6）像以前一样在 TensorFlow 会话中训练模型，但会使用构建器对象保存模型：

```
from tf.saved_model.signature_constants import \
        CLASSIFY_INPUTS
from tf.saved_model.signature_constants import \
        CLASSIFY_OUTPUT_CLASSES
from tf.saved_model.signature_constants import \
        CLASSIFY_OUTPUT_SCORES
from tf.saved_model.signature_constants import \
        CLASSIFY_METHOD_NAME
from tf.saved_model.signature_constants import \
        PREDICT_METHOD_NAME
from tf.saved_model.signature_constants import \
        DEFAULT_SERVING_SIGNATURE_DEF_KEY
with tf.Session() as tfs:
    tfs.run(tf.global_variables_initializer())
    for epoch in range(n_epochs):
        epoch_loss = 0.0
        for batch in range(n_batches):
            x_batch, y_batch = mnist.train.next_batch(batch_size)
            feed_dict = {x_p: x_batch, y_p: y_batch}
            _,batch_loss = tfs.run([train_op,loss],
                        feed_dict=feed_dict)
            epoch_loss += batch_loss
        average_loss = epoch_loss / n_batches
        print("epoch: {0:04d}   loss = {1:0.6f}"
            .format(epoch,average_loss))
    feed_dict={x_p: x_test, y_p: y_test}
    accuracy_score = tfs.run(accuracy_op, feed_dict=feed_dict)
    print("accuracy={0:.8f}".format(accuracy_score))

    # 保存模型

    # 保存模型的定义
    builder = tf.saved_model.builder.SavedModelBuilder(model_dir)
    # 构建 signature_def_map
    bti_op = tf.saved_model.utils.build_tensor_info
    bsd_op = tf.saved_model.utils.build_signature_def

    classification_inputs = bti_op(serialized_tf_example)
    classification_outputs_classes = bti_op(prediction_classes)
    classification_outputs_scores = bti_op(values)
    classification_signature = (bsd_op(
        inputs={CLASSIFY_INPUTS: classification_inputs},
        outputs={CLASSIFY_OUTPUT_CLASSES:
                    classification_outputs_classes,
                CLASSIFY_OUTPUT_SCORES:
                    classification_outputs_scores
            },
        method_name=CLASSIFY_METHOD_NAME))
```

```
            tensor_info_x = bti_op(x_p)
            tensor_info_y = bti_op(probs)

            prediction_signature = (bsd_op(
                inputs={'inputs': tensor_info_x},
                outputs={'outputs': tensor_info_y},
                method_name=PREDICT_METHOD_NAME))

            legacy_init_op = tf.group(tf.tables_initializer(),
                name='legacy_init_op')
            builder.add_meta_graph_and_variables(
                tfs, [tf.saved_model.tag_constants.SERVING],
                signature_def_map={
                    'predict_images':prediction_signature,
                    DEFAULT_SERVING_SIGNATURE_DEF_KEY:
                        classification_signature,
                },
                legacy_init_op=legacy_init_op)

            builder.save()
```

下面是保存模型时输出的信息:

```
accuracy=0.92979997
INFO:tensorflow:No assets to save.
INFO:tensorflow:No assets to write.
INFO:tensorflow:SavedModel written to:
b'/home/armando/models/mnist/1/saved_model.pb'
```

接下来，运行 ModelServer 并提供刚刚保存的模型。

11.3.3 使用 TF 服务提供服务模型

要运行 ModelServer，请执行以下命令:

```
$ tensorflow_model_server --model_name=mnist --
model_base_path=/home/armando/models/mnist
```

服务器开始在端口 8500 上提供模型:

```
I tensorflow_serving/model_servers/main.cc:147] Building single
TensorFlow model file config: model_name: mnist model_base_path:
/home/armando/models/mnist
I tensorflow_serving/model_servers/server_core.cc:441] Adding/
updating models.
I tensorflow_serving/model_servers/server_core.cc:492] (Re-)
adding model: mnist
I tensorflow_serving/core/basic_manager.cc:705] Successfully
reserved resources to load servable {name: mnist version: 1}
I tensorflow_serving/core/loader_harness.cc:66] Approving load
for servable version {name: mnist version: 1}
I tensorflow_serving/core/loader_harness.cc:74] Loading servable
version {name: mnist version: 1}
I
```

```
external/org_tensorflow/tensorflow/contrib/session_bundle/
bundle_ shim.cc:36 0] Attempting to load native
SavedModelBundle in bundle-shim from: /home/armando/
models/mnist/1
I external/org_tensorflow/tensorflow/cc/saved_model/loader.cc:236]
Loading SavedModel from: /home/armando/models/mnist/1
I
external/org_tensorflow/tensorflow/core/platform/cpu_feature_guard.cc:137]
Your CPU supports instructions that this TensorFlow binary was not compiled
to use: AVX2 FMA
I external/org_tensorflow/tensorflow/cc/saved_model/loader.cc:155]
Restoring SavedModel bundle.
I external/org_tensorflow/tensorflow/cc/saved_model/loader.cc:190] Running
LegacyInitOp on SavedModel bundle.
I external/org_tensorflow/tensorflow/cc/saved_model/loader.cc:284] Loading
SavedModel: success. Took 29853 microseconds.
I tensorflow_serving/core/loader_harness.cc:86] Successfully loaded
servable version {name: mnist version: 1}
E1121 ev_epoll1_linux.c:1051] grpc epoll fd: 3
I tensorflow_serving/model_servers/main.cc:288] Running ModelServer at
0.0.0.0:8500 ...
```

可通过调用图像分类模型测试服务器，具体操作可参见 Jupyter 笔记本 ch-11c_TF_Serving_MNIST。

这个笔记本文件的前两个单元提供了来自 TensorFlow 官方示例的客户端测试函数。为了调用 ModelServer，这里修改了该示例，使其在函数签名中能发送 'input' 和接收 'output'。

在笔记本文件中的第三个单元执行了客户端测试函数，具体代码如下：

```
error_rate = do_inference(hostport='0.0.0.0:8500',
                          work_dir='/home/armando/datasets/mnist',
                          concurrency=1,
                          num_tests=100)
print('\nInference error rate: %s%%' % (error_rate * 100))
```

可以得到差不多 7% 的错误率！（不同用户可能会得到不同的值）：

```
Extracting /home/armando/datasets/mnist/train-images-idx3-ubyte.gz
Extracting /home/armando/datasets/mnist/train-labels-idx1-ubyte.gz
Extracting /home/armando/datasets/mnist/t10k-images-idx3-ubyte.gz
Extracting /home/armando/datasets/mnist/t10k-labels-idx1-ubyte.gz

..........................................................
..........................................................
Inference error rate: 7.0%
```

11.4 在 Docker 容器中提供 TF 服务

Docker 是一个用于在容器中打包和部署应用程序的平台。如果您还不了解 Docker 容器，请访问以下链接：https://www.docker.com/what-container。

还可以在 Docker 容器中安装和运行 TensorFlow 服务，本节中提供的 Ubuntu 16.04 的说明源于 TensorFlow 官方网站上的链接：

- https://www.tensorflow.org/serving/serving_inception
- https://www.tensorflow.org/serving/serving_basic

11.4.1 安装 Docker

按如下方式安装 Docker：

1）首先，删除之前安装的 Docker：

```
$ sudo apt-get remove docker docker-engine docker.io
```

2）安装必备软件：

```
$ sudo apt-get install \
    apt-transport-https \
    ca-certificates \
    curl \
    software-properties-common
```

3）为 Docker 资料库 (repository) 添加 GPG 密钥：

```
$ curl -fsSL https://download.docker.com/linux/ubuntu/gpg | sudo apt-key add -
```

4）添加 Docker 资料库：

```
$ sudo add-apt-repository \
    "deb [arch=amd64] https://download.docker.com/linux/ubuntu \
    $(lsb_release -cs) \
    stable"
```

5）安装社区版 Docker：

```
$ sudo apt-get update && sudo apt-get install docker-ce
```

6）安装成功后，将 Docker 添加为系统服务：

```
$ sudo systemctl enable docker
```

7）要以非 root 用户身份运行 Docker 或不使用 sudo，请添加 docker 组：

```
$ sudo groupadd docker
```

8）将用户添加到 docker 组：

```
$ sudo usermod -aG docker $USER
```

9）现在注销并再次登录，以便组成员身份生效。登录后，运行以下命令测试 Docker 安装：

```
$ docker run --name hello-world-container hello-world
```

可以看到类似下面的输出结果：

```
Unable to find image 'hello-world:latest' locally
latest: Pulling from library/hello-world
ca4f61b1923c: Already exists
Digest:
sha256:be0cd392e45be79ffeffa6b05338b98ebb16c87b255f48e297ec7f98e123905c
```

```
Status: Downloaded newer image for hello-world:latest

Hello from Docker!
This message shows that your installation appears to be working correctly.

To generate this message, Docker took the following steps:
 1. The Docker client contacted the Docker daemon.
 2. The Docker daemon pulled the "hello-world" image from the Docker Hub.
    (amd64)
 3. The Docker daemon created a new container from that image which runs
the
    executable that produces the output you are currently reading.
 4. The Docker daemon streamed that output to the Docker client, which sent
it
    to your terminal.

To try something more ambitious, you can run an Ubuntu container with:
 $ docker run -it ubuntu bash
Share images, automate workflows, and more with a free Docker ID:
 https://cloud.docker.com/

For more examples and ideas, visit:
 https://docs.docker.com/engine/userguide
```

Docker 安装成功后，需要为 TensorFlow 服务构建一个 Docker 镜像。

11.4.2 为 TF 服务构建 Docker 镜像

生成 Docker 镜像的操作如下：

1）使用以下内容创建名为 dockerfile 的文件：

```
FROM ubuntu:16.04
MAINTAINER Armando Fandango <armando@geekysalsero.com>

RUN apt-get update && apt-get install -y \
    build-essential \
    curl \
    git \
    libfreetype6-dev \
    libpng12-dev \
    libzmq3-dev \
    mlocate \
    pkg-config \
    python-dev \
    python-numpy \
    python-pip \
    software-properties-common \
    swig \
    zip \
    zlib1g-dev \
    libcurl3-dev \
    openjdk-8-jdk\
    openjdk-8-jre-headless \
    wget \
```

```
    && \
    apt-get clean && \
    rm -rf /var/lib/apt/lists/*

RUN echo "deb [arch=amd64]
http://storage.googleapis.com/tensorflow-serving-apt stable
tensorflow-model-server tensorflow-model-server-universal" \
    | tee /etc/apt/sources.list.d/tensorflow-serving.list
RUN curl
https://storage.googleapis.com/tensorflow-serving-apt/tensorflow-se
rving.release.pub.gpg \
    | apt-key add -

RUN apt-get update && apt-get install -y \
    tensorflow-model-server

RUN pip install --upgrade pip
RUN pip install mock grpcio tensorflow tensorflow-serving-api

CMD ["/bin/bash"]
```

2）运行以下命令从上述 dockerfile 中构建 Docker 镜像：

```
$ docker build --pull -t $USER/tensorflow_serving -f dockerfile .
```

3）创建镜像需要一段时间，当看到类似以下内容时，表示镜像构建成功：

```
Removing intermediate container 1d8e757d96e0
Successfully built 0f95ddba4362
Successfully tagged armando/tensorflow_serving:latest
```

4）运行以下命令以启动容器：

```
$ docker run --name=mnist_container -it $USER/tensorflow_serving
```

5）当看到以下提示时，表示已经登录到容器中了：

```
root@244ea14efb8f:/#
```

6）将 cd 命令切换到 home 目录。

7）在 home 目录中，输入以下命令检查 TensorFlow 的服务是否已经启动。这里使用代码中的示例演示，但大家可以查看自己的 Git 资料库来运行自己的模型。

```
$ git clone --recurse-submodules
https://github.com/tensorflow/serving
```

克隆资料库后，就可以构建、训练和保存 MNIST 模型了。

8）使用以下命令删除临时目录（如果尚未删除）：

```
$ rm -rf /tmp/mnist_model
```

9）运行以下命令构建、训练和保存 MNIST 模型：

```
$ python serving/tensorflow_serving/example/mnist_saved_model.py
/tmp/mnist_model
```

将看到类似于下面的输出结果：

```
Training model...
Successfully downloaded train-images-idx3-ubyte.gz 9912422 bytes.
Extracting /tmp/train-images-idx3-ubyte.gz
Successfully downloaded train-labels-idx1-ubyte.gz 28881 bytes.
Extracting /tmp/train-labels-idx1-ubyte.gz
Successfully downloaded t10k-images-idx3-ubyte.gz 1648877 bytes.
Extracting /tmp/t10k-images-idx3-ubyte.gz
Successfully downloaded t10k-labels-idx1-ubyte.gz 4542 bytes.
Extracting /tmp/t10k-labels-idx1-ubyte.gz
2017-11-22 01:09:38.165391: I
tensorflow/core/platform/cpu_feature_guard.cc:137] Your CPU
supports instructions that this TensorFlow binary was not compiled
to use: SSE4.1 SSE4.2 AVX AVX2 FMA
training accuracy 0.9092
Done training!
Exporting trained model to /tmp/mnist_model/1
Done exporting!
```

10）按"Ctrl + P"和"Ctrl + Q"键从 Docker 镜像中分离。

11）提交新映像的变化，并使用以下命令停止容器：

```
$ docker commit mnist_container $USER/mnist_serving
$ docker stop mnist_container
```

12）现在可以通过以下命令随时运行该容器：

```
$ docker run --name=mnist_container -it $USER/mnist_serving
```

13）删除为保存镜像而构建的临时 MNIST 容器：

```
$ docker rm mnist_container
```

11.4.3 在 Docker 容器中提供模型

可通过下面的步骤实现在容器中提供模型：

1）启动上一节中构建的 MNIST 容器：

```
$ docker run --name=mnist_container -it $USER/mnist_serving
```

2）使用 cd 命令转到 Home 目录下。

3）使用以下命令运行 ModelServer：

```
$ tensorflow_model_server  --model_name=mnist --
model_base_path=/tmp/mnist_model/ &> mnist_log &
```

4）使用客户端检查模型的预测结果：

```
$ python serving/tensorflow_serving/example/mnist_client.py --
num_tests=100 --server=localhost:8500
```

5）看到错误率为 7%，这与 Jupyter 笔记本示例执行的一样：

```
Extracting /tmp/train-images-idx3-ubyte.gz
Extracting /tmp/train-labels-idx1-ubyte.gz
Extracting /tmp/t10k-images-idx3-ubyte.gz
Extracting /tmp/t10k-labels-idx1-ubyte.gz
```

```
............................................................
..............................
Inference error rate: 7.0%
```

这里已经构建了一个 Docker 镜像，并在 Docker 镜像中为模型提供了服务，执行 exit 命令可退出容器。

11.5 基于 Kubernetes 的 TF 服务

Kubernetes 是一个开源系统，用于自动化、容器化应用程序的部署、扩展和管理（参考 https://kubernets.io）。在生产环境中使用 Kubernetes 集群可将 TensorFlow 模型扩展为数百个或数千个 TF 服务。Kubernetes 集群可以在所有流行的公共云上运行，例如 GCP、AWS、Azure，以及本地私有云。因此，下面先介绍如何安装 Kubernetes，然后在 Kubernetes 集群上部署 MNIST 模型。

11.5.1 安装 Kubernetes

在 Ubuntu 16.04 上以单节点本地集群安装 Kubernetes 的步骤如下：

1）安装 LXD 和 Docker，这是在本地安装 Kubernetes 的先决条件。LXD 是与 Linux 容器一起使用的容器管理器。在上一节中已经学习了如何安装 Docker。可运行以下命令安装 LXD：

```
$ sudo snap install lxd
lxd 2.19 from 'canonical' installed
```

2）初始化 lxd 并创建虚拟网络：

```
$ sudo /snap/bin/lxd init --auto
LXD has been successfully configured.

$ sudo /snap/bin/lxc network create lxdbr0 ipv4.address=auto ipv4.nat=true ipv6.address=none ipv6.nat=false
If this is your first time using LXD, you should also run: lxd init
 To start your first container, try: lxc launch ubuntu:16.04

Network lxdbr0 created
```

3）将用户添加到 lxd 组：

```
$ sudo usermod -a -G lxd $(whoami)
```

4）安装 conjure-up 并重启机器：

```
$ sudo snap install conjure-up --classic
conjure-up 2.3.1 from 'canonical' installed
```

5）执行 conjure-up 以安装 Kubernetes：

```
$ conjure-up kubernetes
```

6）从 spell 列表中选择 Kubernetes Core。

7）从可用云列表中选择 localhost。

8）从网络列表中选择 lxbr0 bridge。

9）为这两个选项提供 sudo 密码：Download the kubectl and kubefedclient programs to your local host。

10）在下一个屏幕中，会要求选择要安装的应用程序。安装剩下的所有 5 个应用程序。当安装期间的最终屏幕如图 11-1 所示时，表示 Kubernetes 集群准备进行工作。

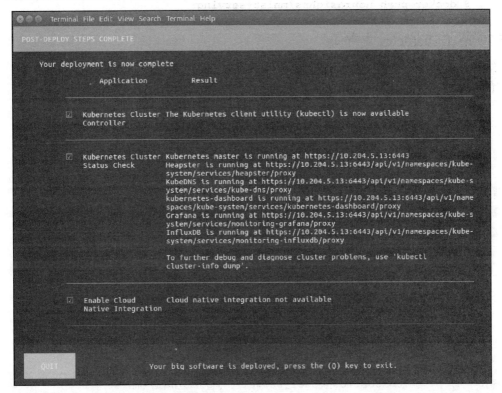

图 11-1

如果在安装时遇到问题，可以先查阅以下链接的文档，或者在互联网上搜索相关的帮助信息：

https://kubernetes.io/docs/getting-started-guides/ubuntu/local/。

https://kubernetes.io/docs/getting-started-guides/ubuntu/。

https://tutorials.ubuntu.com/tutorial/install-kubernetes-with-conjure-up。

11.5.2　将 Docker 镜像上传到 dockerhub

将 Docker 镜像上传到 dockerhub 的步骤如下：

1）如果还没有 dockerhub 账户，需要创建一个。

2）使用以下命令登录 dockerhub 账户：

```
$ docker login --username=<username>
```

3）使用您在 dockerhub 上创建的资料库标记 MNIST 镜像，例如，创建 neurasights/mnist-serving：

```
$ docker tag $USER/mnist_serving neurasights/mnist-serving
```

4）将标记的镜像推送到 dockerhub 账户：

```
$ docker push neurasights/mnist-serving
```

11.5.3　在 Kubernetes 中部署

继续在 Kubernotes 中进行部署，其步骤如下：

1）使用以下代码创建 mnist.yaml 文件：

```yaml
apiVersion: extensions/v1beta1
kind: Deployment
metadata:
  name: mnist-deployment
spec:
  replicas: 3
  template:
    metadata:
      labels:
        app: mnist-server
    spec:
      containers:
      - name: mnist-container
        image: neurasights/mnist-serving
        command:
        - /bin/sh
        args:
        - -c
        - tensorflow_model_server --model_name=mnist --model_base_path=/tmp/mnist_model
        ports:
        - containerPort: 8500
---
apiVersion: v1
kind: Service
metadata:
  labels:
    run: mnist-service
  name: mnist-service
spec:
  ports:
  - port: 8500
    targetPort: 8500
  selector:
    app: mnist-server
#  类型: LoadBalancer
```

第 11 章 使用 TF 服务提供生成环境下的 TensorFlow 模型

 TIP 对于上面这个 mnist.yaml，如果在 AWS 或 GCP 云中运行，需要取消 LoadBalancer 行的注释。因为我们只在本地集群的一个节点上运行，所以没有外部 LoadBalancer。

2）创建 Kubernetes 部署和服务：

```
$ kubectl create -f mnist.yaml
deployment "mnist-deployment" created
service "mnist-service" created
```

3）检查部署、pod 和服务：

```
$ kubectl get deployments
NAME               DESIRED   CURRENT   UP-TO-DATE   AVAILABLE   AGE
mnist-deployment   3         3         3            0           1m

$ kubectl get pods
NAME                                      READY     STATUS              RESTARTS   AGE
default-http-backend-bbchw                1/1       Running             3          9d
mnist-deployment-554f4b674b-pwk8z         0/1       ContainerCreating   0          1m
mnist-deployment-554f4b674b-vn6sd         0/1       ContainerCreating   0          1m
mnist-deployment-554f4b674b-zt4xt         0/1       ContainerCreating   0          1m
nginx-ingress-controller-724n5            1/1       Running             2          9d

$ kubectl get services
NAME                   TYPE           CLUSTER-IP       EXTERNAL-IP   PORT(S)          AGE
default-http-backend   ClusterIP      10.152.183.223   <none>        80/TCP           9d
kubernetes             ClusterIP      10.152.183.1     <none>        443/TCP          9d
mnist-service          LoadBalancer   10.152.183.66    <pending>     8500:32414/TCP   1m

$ kubectl describe service mnist-service
Name:                   mnist-service
Namespace:              default
Labels:                 run=mnist-service
Annotations:            <none>
Selector:               app=mnist-server
Type:                   LoadBalancer
IP:                     10.152.183.66
Port:                   <unset>  8500/TCP
TargetPort:             8500/TCP
NodePort:               <unset>  32414/TCP
Endpoints:
```

```
10.1.43.122:8500,10.1.43.123:8500,10.1.43.124:8500
Session Affinity:              None
External Traffic Policy:       Cluster
Events:                        <none>
```

4）等到所有 pod 都处于运行状态：

```
$ kubectl get pods
NAME                                         READY     STATUS      RESTARTS
AGE
default-http-backend-bbchw                   1/1       Running     3
9d
mnist-deployment-554f4b674b-pwk8z            1/1       Running     0
3m
mnist-deployment-554f4b674b-vn6sd            1/1       Running     0
3m
mnist-deployment-554f4b674b-zt4xt            1/1       Running     0
3m
nginx-ingress-controller-724n5               1/1       Running     2
9d
```

5）检查其中一个 pod 日志，可以看到如下内容：

```
$ kubectl logs mnist-deployment-59dfc5df64-g7prf
I tensorflow_serving/model_servers/main.cc:147] Building single
TensorFlow model file config: model_name: mnist model_base_path:
/tmp/mnist_model
I tensorflow_serving/model_servers/server_core.cc:441]
Adding/updating models.
I tensorflow_serving/model_servers/server_core.cc:492] (Re-)adding
model: mnist
I tensorflow_serving/core/basic_manager.cc:705] Successfully
reserved resources to load servable {name: mnist version: 1}
I tensorflow_serving/core/loader_harness.cc:66] Approving load for
servable version {name: mnist version: 1}
I tensorflow_serving/core/loader_harness.cc:74] Loading servable
version {name: mnist version: 1}
I
external/org_tensorflow/tensorflow/contrib/session_bundle/bundle_sh
im.cc:360] Attempting to load native SavedModelBundle in bundle-
shim from: /tmp/mnist_model/1
I external/org_tensorflow/tensorflow/cc/saved_model/loader.cc:236]
Loading SavedModel from: /tmp/mnist_model/1
I
external/org_tensorflow/tensorflow/core/platform/cpu_feature_guard.
cc:137] Your CPU supports instructions that this TensorFlow binary
was not compiled to use: AVX2 FMA
I external/org_tensorflow/tensorflow/cc/saved_model/loader.cc:155]
Restoring SavedModel bundle.
I external/org_tensorflow/tensorflow/cc/saved_model/loader.cc:190]
Running LegacyInitOp on SavedModel bundle.
I external/org_tensorflow/tensorflow/cc/saved_model/loader.cc:284]
Loading SavedModel: success. Took 45319 microseconds.
I tensorflow_serving/core/loader_harness.cc:86] Successfully loaded
```

```
servable version {name: mnist version: 1}
E1122 12:18:04.566415410 6 ev_epoll1_linux.c:1051] grpc epoll fd: 3
I tensorflow_serving/model_servers/main.cc:288] Running ModelServer
    at 0.0.0.0:8500 ...
```

6）还可以使用以下命令查看 UI 控制台：

```
$ kubectl proxy xdg-open http://localhost:8001/ui
```

Kubernetes UI 控制台如图 11-2 和图 11-3 所示。

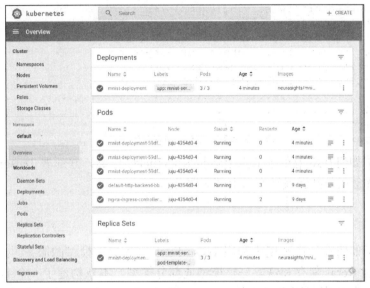

图 11-2

图 11-3

由于在本地集群的单个节点上运行，因此服务仅在集群中公开，无法从外部访问。登录刚刚实例化的三个 pod 中的一个：

```
$ kubectl exec -it mnist-deployment-59dfc5df64-bb24q -- /bin/bash
```

切换到主目录并运行 MNIST 客户端测试服务：

```
$ kubectl exec -it mnist-deployment-59dfc5df64-bb24q -- /bin/bash
root@mnist-deployment-59dfc5df64-bb24q:/# cd
root@mnist-deployment-59dfc5df64-bb24q:~# python
serving/tensorflow_serving/example/mnist_client.py --num_tests=100 --
server=10.152.183.67:8500
Extracting /tmp/train-images-idx3-ubyte.gz
Extracting /tmp/train-labels-idx1-ubyte.gz
Extracting /tmp/t10k-images-idx3-ubyte.gz
Extracting /tmp/t10k-labels-idx1-ubyte.gz
..........................................................................
...............................
Inference error rate: 7.0%
root@mnist-deployment-59dfc5df64-bb24q:~#
```

这里学习了如何在本地单个节点上运行 Kubernetes 集群，并部署 TensorFlow 服务，可以使用相同的概念在公共云或私有云上部署服务。

11.6 总结

本章介绍了如何在生产环境中利用 TensorFlow 服务提供模型的服务，还介绍了如何使用 TensorFlow 和 Keras 保存和恢复完整模型或部分模型。本章构建了一个 Docker 容器，从官方 TensorFlow 服务资料库中得到基于 Docker 容器中的 MNIST 示例代码，并运行了这些代码。本章还安装了一个本地 Kubernetes 集群，并部署了 MNIST 模型，用于在 Kubernetes pod 中运行 TensorFlow 服务。鼓励读者在这些例子的基础上进一步尝试，比如尝试提供不同的模型。TF 服务文档描述了各种选项，并提供了其他信息，可以进一步研究这方面的内容。

在接下来的章节中，将通过使用迁移学习实现高级模型。TensorFlow 资料库中提供的预训练模型是练习通过使用 TF 服务提供模型的最佳选择。这里介绍了如何使用 Ubuntu 软件包安装 TF 服务，但为了优化生产环境，有可能需要从源代码构建 TF 服务。

第 12 章
迁移学习模型和预训练模型

简单来说，迁移学习是指在得到能预测一种类别的预训练模型后，直接使用该模型或仅重新训练该模型的一小部分，便可预测另一种类别。例如，可以采用预训练模型识别猫的类型，然后仅针对狗的类型重新训练模型的一小部分，就可使用新模型预测狗的类型。

如果没有迁移学习，在大型数据集上训练一个巨大的模型需要几天甚至几个月。然而，通过迁移学习，使用预先训练的模型，仅需要训练最后几层，而不需要从头开始训练，从而可以节省大量时间。

在没有大量数据集的情况下，迁移学习也很有用。在小型数据集上训练的模型可能无法检测在大型数据集上训练得到的模型特征。但通过迁移学习，即使数据集较小，也可以获得更好的模型。

本章将采用预先训练的模型对新物体进行训练。本章将展示基于图像的预训练模型的示例，并将其应用于图像分类问题。可以尝试找到其他预先训练的模型，并应用于不同的问题，如目标检测、文本生成或机器翻译。本章将介绍以下主题：

- ImageNet 数据集；
- 重新训练或微调模型；
- COCO 动物数据集和预处理图像；
- 使用 TensorFlow 中预先训练的 VGG16 进行图像分类；
- 将 TensorFlow 中的图像预处理用于预先训练的 VGG16；
- 使用 TensorFlow 中重新训练的 VGG16 进行图像分类；
- 使用 Keras 中预先训练的 VGG16 进行图像分类；
- 使用 Keras 中重新训练的 VGG16 进行图像分类；
- 使用 TensorFlow 中 Inception v3 进行图像分类；
- 使用 TensorFlow 中重新训练的 Inception v3 进行图像分类。

12.1 ImageNet 数据集

ImageNet 是根据 WordNet 层次结构组织的图像数据集。WordNet 中的每个有意义的概念（可能由多个单词或单词短语描述）称为同义词集（synonym set 或 synset）（参见 http://image-net.org）。

ImageNet 有大约 100 K 个同义词集，平均每个同义词集约有 1 000 个人工注释图像。ImageNet 仅存储对图像的引用，而图像存储在 Internet 上的原始位置。在深度学习论文中，ImageNet-1K 是指作为 ImageNet 一部分的**大规模视觉识别挑战（ILSVRC）**所发布的数据集，它将数据集分为 1 000 个类别。

可以在以下 URL 中找到 1 000 个挑战类别：

http://image-net.org/challenges/LSVRC/2017/browse-synsets。
http://image-net.org/challenges/LSVRC/2016/browse-synsets。
http://image-net.org/challenges/LSVRC/2015/browse-synsets。
http://image-net.org/challenges/LSVRC/2014/browse-synsets。
http://image-net.org/challenges/LSVRC/2013/browse-synsets。
http://image-net.org/challenges/LSVRC/2012/browse-synsets。
http://image-net.org/challenges/LSVRC/2011/browse-synsets。
http://image-net.org/challenges/LSVRC/2010/browse-synsets。

编写一个自定义函数从 Google 下载 ImageNet 的类标签：

```
def build_id2label(self):
    base_url = 'https://raw.githubusercontent.com/tensorflow/models/master/research/inception/inception/data/'
    synset_url = '{}/imagenet_lsvrc_2015_synsets.txt'.format(base_url)
    synset_to_human_url = '{}/imagenet_metadata.txt'.format(base_url)

    filename, _ = urllib.request.urlretrieve(synset_url)
    synset_list = [s.strip() for s in open(filename).readlines()]
    num_synsets_in_ilsvrc = len(synset_list)
    assert num_synsets_in_ilsvrc == 1000

    filename, _ = urllib.request.urlretrieve(synset_to_human_url)
    synset_to_human_list = open(filename).readlines()
    num_synsets_in_all_imagenet = len(synset_to_human_list)
    assert num_synsets_in_all_imagenet == 21842

    synset2name = {}
    for s in synset_to_human_list:
        parts = s.strip().split('\t')
        assert len(parts) == 2
        synset = parts[0]
        name = parts[1]
        synset2name[synset] = name

    if self.n_classes == 1001:
        id2label={0:'empty'}
        id=1
    else:
        id2label = {}
        id=0
```

```
for synset in synset_list:
    label = synset2name[synset]
    id2label[id] = label
    id += 1

return id2label
```

按如下方式将这些标签加载到 Jupyter 笔记本中：

```
### 加载ImageNet数据集的标签
from datasetslib.imagenet import imageNet
inet = imageNet()
inet.load_data(n_classes=1000)
#对于Inception模型，n_classes为1001；对于VGG模型，n_classes为1000
```

表 12-1 为由 ImageNet-1K 数据集训练得到的预训练图像分类模型。

表 12-1

模型名	Top-1 分类精度	Top-5 分类精度	Top-5 错误率	原始论文的链接
AlexNet			15.3%	https://www.cs.toronto.edu/~fritz/absps/imagenet.pdf
Inception also known as Inception V1	69.8	89.6	6.67%	https://arxiv.org/abs/1409.4842
BN-Inception-v2 also known as Inception V2	73.9	91.8	4.9%	https://arxiv.org/abs/1502.03167
Inception V3	78.0	93.9	3.46%	https://arxiv.org/abs/1512.00567
Inception V4	80.2	95.2		http://arxiv.org/abs/1602.07261
Inception-Resenet-V2	80.4	95.2		http://arxiv.org/abs/1602.07261
VGG16	71.5	89.8	7.4%	https://arxiv.org/abs/1409.1556
VGG19	71.1	89.8	7.3%	https://arxiv.org/abs/1409.1556
ResNet V1 50	75.2	92.2	7.24%	https://arxiv.org/abs/1512.03385
ResNet V1 101	76.4	92.9		https://arxiv.org/abs/1512.03385
ResNet V1 152	76.8	93.3		https://arxiv.org/abs/1512.03385
ResNet V2 50	75.6	92.8		https://arxiv.org/abs/1603.05027
ResNet V2 101	77.0	93.7		https://arxiv.org/abs/1603.05027
ResNet V2 152	77.8	94.1		https://arxiv.org/abs/1603.05027
ResNet V2 200	79.9	95.2		https://arxiv.org/abs/1603.05027
Xception	79.0	94.5		https://arxiv.org/abs/1610.02357
MobileNet V1 versions	41.3～70.7	66.2～89.5		https://arxiv.org/pdf/1704.04861.pdf

在表 12-1 中，Top-1 和 Top-5 指标指的是模型在 ImageNet 验证数据集上的性能。

Google Research 最近提出了一种名为 MobileNets 的新型深度神经网络，MobileNets 采用移动设备优先策略开发，为了降低资源的使用，牺牲一些精度，该模型能提供低功耗、低延迟模型，以便能在移动和嵌入式设备上提供更好的体验。Google 为 MobileNet 模型提供了 16 个预先训练好的检查点文件，每个模型提供不同数量的参数和**乘积累加运算**（Multiply-Accumulates，MAC）。MAC 和参数越高，资源使用和延迟就越高。因此，可以在更高的准确性与更高的资源使用/延迟之间进行选择（见表 12-2）。

表 12-2

模型的检查点文件	MAC（以百万为单位）	参数（以百万为单位）	Top-1 分类精度	Top-5 分类精度
MobileNet_v1_1.0_224	569	4.24	70.7	89.5
MobileNet_v1_1.0_192	418	4.24	69.3	88.9
MobileNet_v1_1.0_160	291	4.24	67.2	87.5
MobileNet_v1_1.0_128	186	4.24	64.1	85.3
MobileNet_v1_0.75_224	317	2.59	68.4	88.2
MobileNet_v1_0.75_192	233	2.59	67.4	87.3
MobileNet_v1_0.75_160	162	2.59	65.2	86.1
MobileNet_v1_0.75_128	104	2.59	61.8	83.6
MobileNet_v1_0.50_224	150	1.34	64.0	85.4
MobileNet_v1_0.50_192	110	1.34	62.1	84.0
MobileNet_v1_0.50_160	77	1.34	59.9	82.5
MobileNet_v1_0.50_128	49	1.34	56.2	79.6
MobileNet_v1_0.25_224	41	0.47	50.6	75.0
MobileNet_v1_0.25_192	34	0.47	49.0	73.6
MobileNet_v1_0.25_160	21	0.47	46.0	70.7
MobileNet_v1_0.25_128	14	0.47	41.3	66.2

> 有关 MobileNets 的更多信息，请访问以下资源：
> https://research.googleblog.com/2017/06/mobilenets-open-source-models-for.html.
> https://github.com/tensorflow/models/blob/master/research/slim/nets/mobilenet_v1.md.
> https://arxiv.org/pdf/1704.04861.pdf.

12.2 重新训练或微调模型

在 ImageNet 这种大型且多样化的数据集上训练的模型能够检测和捕获一些通用特征，如曲线、边缘和形状。其中一些特征很容易适用其他类型的数据集。因此，在迁移学习中，对于新的数据集，会采用这样的通用模型，并使用以下一些技术微调或重新训练模型：

- **替换最后一层**：最常见的做法是删除最后一层并添加与新数据集相关的分类网络层。例如，ImageNet 模型使用 1 000 个类别进行训练，但 COCO 动物数据集只有 8 个类别，因此可删除 softmax 层，该层使用 softmax 函数得到 1 000 个类别的概率，可使用新的网络层得到 8 个类别的概率。通常，当新数据集几乎与训练模型的数据集类似时，可使用此技术，在这种情况下只需要重新训练最后一层即可。

- **固定前几层**：另一种常见做法是固定前几层，以便仅使用新数据集更新最后未固定层的权重。本章会给出一个固定前 15 层，只重新训练最后 10 层的例子。通常，当新数据集与训练模型的数据集非常不一样时才使用该技术，这种方法只需要训练最后几层。

- **调整超参数**：还可以在重新训练之前调整超参数，例如更改学习速率或尝试不同的损失函数、不同的优化器。

TensorFlow 和 Keras 都提供预训练模型。

下面将通过 TensorFlow Slim 演示示例。可以在文件夹 tensorflow / models / research / slim / nets 中找到几个预先训练的模型（在编写本书时至少是这样的目录结构）。本章将使用 TensorFlow Slim 预先训练的模型进行实例化，然后从下载的检查点文件加载权重。加载的模型将用于对新数据集进行预测，然后重新训练模型微调预测。

本章还将通过 keras.applications 模块中提供的 Keras 预训练模型演示迁移学习。TensorFlow 有大约 20 多个预训练模型，但 keras.appplications 只有以下 7 个预训练模型：

- Xception - https://keras.io/applications/#xception
- VGG16 - https://keras.io/applications/#vgg16
- VGG19 - https://keras.io/applications/#vgg19
- ResNet50 - https://keras.io/applications/#resnet50
- InceptionV3 - https://keras.io/applications/#inceptionv3
- InceptionResNetV2 - https://keras.io/applications/#inceptionresnetv2
- MobileNet - https://keras.io/applications/#mobilenet

12.3　COCO 动物数据集和预处理图像

本次示例将使用 COCO 动物数据集，这是 COCO 数据集的一小部分，由斯坦福大学的研究人员提供，链接如下：http://cs231n.stanford.edu/coco-animals.zip。COCO 动物数据集有 800 幅训练图像和 8 个动物类别的 200 幅测试图像，这些图像中的动物分别为熊、鸟、猫、狗、长颈鹿、马、绵羊和斑马。下载图像并对 VGG16 和 Inception 模型进行预处理。

对于 VGG 模型，图像大小为 224×224，预处理步骤如下：

1）使用与 TensorFlow 的 tf.image.resize_image_with_crop_or_pad 函数类似的函数将图像大小调整为 224×224。函数实现如下：

```
def resize_image(self,in_image:PIL.Image, new_width,
    new_height, crop_or_pad=True):
    img = in_image
    if crop_or_pad:
        half_width = img.size[0] // 2
        half_height = img.size[1] // 2
        half_new_width = new_width // 2
        half_new_height = new_height // 2
        img = img.crop((half_width-half_new_width,
                        half_height-half_new_height,
                        half_width+half_new_width,
                        half_height+half_new_height
                      ))
    img = img.resize(size=(new_width, new_height))

    return img
```

2）调整大小后，将图像从 PIL.Image 转换为 NumPy Array 并检查图像是否有深度通道，执行这样的检查是因为数据集中有些图像为灰度图像。

```
img = self.pil_to_nparray(img)
if len(img.shape)==2:
    # 灰度图像或无通道，然后添加三个通道
    h=img.shape[0]
    w=img.shape[1]
    img = np.dstack([img]*3)
```

3）为了让数据居中，需从图像中减去 VGG 数据集平均值。将新训练图像居中是为了让这些特征具有与初始数据类似的范围。让这些特征在相似范围内，可以确保重新训练期间的梯度不会变得太高或太低。而且通过使数据居中，学习过程会变得更快，因为以零均值为中心的每个通道的梯度会变得均匀。

```
means = np.array([[[123.68, 116.78, 103.94]]]) #形状为[1, 1, 3]
img = img - means
```

完整的预处理函数如下：

```
def preprocess_for_vgg(self,incoming, height, width):
    if isinstance(incoming, six.string_types):
        img = self.load_image(incoming)
    else:
        img=incoming
    img_size = vgg.vgg_16.default_image_size
    height = img_size
    width = img_size
    img = self.resize_image(img,height,width)
    img = self.pil_to_nparray(img)
    if len(img.shape)==2:
        # 灰度图像或无通道，然后添加三个通道
        h=img.shape[0]
        w=img.shape[1]
        img = np.dstack([img]*3)

    means = np.array([[[123.68, 116.78, 103.94]]]) #形状为[1, 1, 3]
    try:
        img = img - means
    except Exception as ex:
        print('Error preprocessing ',incoming)
        print(ex)

    return img
```

对于 Inception 模型，图像大小为 299×299，预处理步骤如下：

1）使用与 TensorFlow 的 tf.image.resize_image_with_crop_or_pad 函数类似的函数将图像大小调整为 299×299。正如之前在 VGG 预处理步骤中实现的函数一样。

2）然后使用以下代码将图像缩放到（-1，+1）之间：

```
img = ((img/255.0) - 0.5) * 2.0
```

完整的预处理函数如下：

```python
def preprocess_for_inception(self,incoming):
    img_size = inception.inception_v3.default_image_size
    height = img_size
    width = img_size
    if isinstance(incoming, six.string_types):
        img = self.load_image(incoming)
    else:
        img=incoming
    img = self.resize_image(img,height,width)
    img = self.pil_to_nparray(img)
    if len(img.shape)==2:
        # 灰度图像或无通道，然后添加三个通道
        h=img.shape[0]
        w=img.shape[1]
        img = np.dstack([img]*3)
    img = ((img/255.0) - 0.5) * 2.0

    return img
```

加载 COCO 动物数据集：

```python
from datasetslib.coco import coco_animals
coco = coco_animals()
x_train_files, y_train, x_val_files, x_val = coco.load_data()
```

从验证集的每个类别中获取一幅图像来生成列表 x_test，并通过预处理图像生成列表 images_test：

```python
x_test = [x_val_files[25*x] for x in range(8)]
images_test=np.array([coco.preprocess_for_vgg(x) for x in x_test])
```

使用这个 helper 函数显示与图像相关的前 5 个类别的图像和概率：

```python
# helper函数
def disp(images,id2label=None,probs=None,n_top=5,scale=False):
    if scale:
        imgs = np.abs(images + np.array([[[[123.68,
            116.78, 103.94]]]]))/255.0
    else:
        imgs = images
    ids={}
    for j in range(len(images)):
        if scale:
            plt.figure(figsize=(5,5))
            plt.imshow(imgs[j])
        else:
            plt.imshow(imgs[j].astype(np.uint8) )
        plt.show()
        if probs is not None:
            ids[j] = [i[0] for i in sorted(enumerate(-probs[j]),
                    key=lambda x:x[1])]
            for k in range(n_top):
                id = ids[j][k]
                print('Probability {0:1.2f}% of[{1:}]'
                    .format(100*probs[j,id],id2label[id]))
```

在上述函数中有如下一行代码，它是用来恢复预处理的效果，以便能显示原始图像而不是预处理后的图像：

```
imgs = np.abs(images + np.array([[[[123.68, 116.78, 103.94]]]]))/255.0
```

在 Inception 模型下，用于恢复预处理图像的代码如下：

```
imgs = (images / 2.0) + 0.5
```

可以使用以下代码查看测试图像：

```
images=np.array([mpimg.imread(x) for x in x_test])
disp(images)
```

按照 Jupyter 笔记本中的代码查看图像，这些图像看起来都有不同的大小，因此需要输出这些图像的原始尺寸：

```
print([x.shape for x in images])
```

图像维度为：

```
[(640, 425, 3), (373, 500, 3), (367, 640, 3), (427, 640, 3), (428, 640, 3),
 (426, 640, 3), (480, 640, 3), (612, 612, 3)]
```

预处理测试图像并查看图像的大小：

```
images_test=np.array([coco.preprocess_for_vgg(x) for x in x_test])
print(images_test.shape)
```

维度为：

```
(8, 224, 224, 3)
```

在 Inception 的情况下，维度为：

```
(8, 299, 299, 3)
```

初始化的预处理图像不可见，但可以输出 VGG 的预处理图像，以了解图像的内容（见图 12-1）：

```
disp(images_test)
```

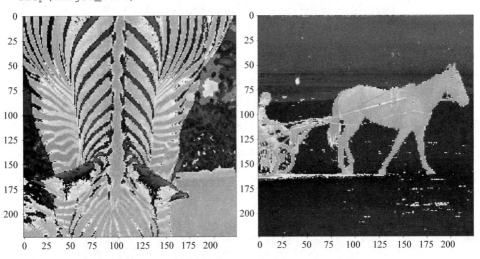

图 12-1

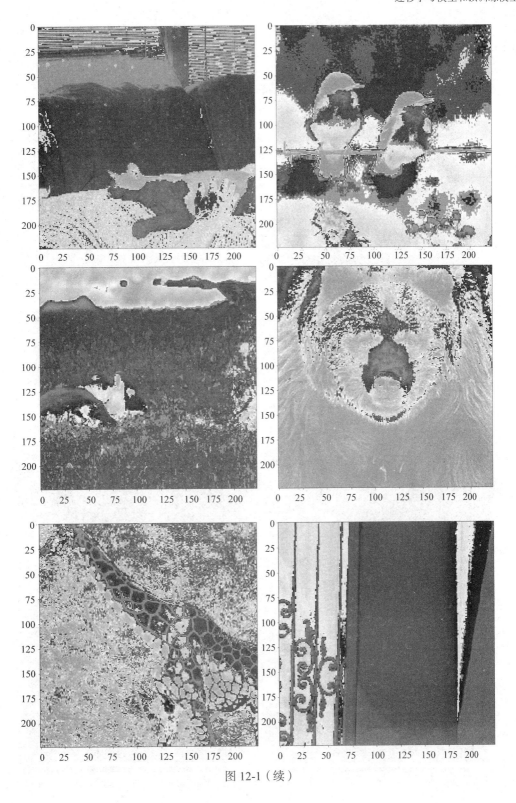

图 12-1（续）

图像实际上被裁剪了,可以看到对图像保持裁剪的同时,恢复预处理图像的样子(见图 12-2)。

图 12-2

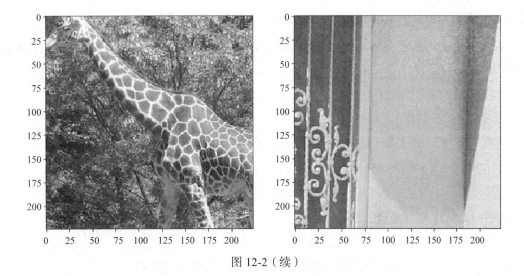

图 12-2（续）

现在已经得到 ImageNet 的类标签，并加载了 COCO 图像数据集和相应的标签，下面介绍迁移学习示例。

12.4 TensorFlow 中的 VGG16

 可以按照 Jupyter 笔记本 ch-12a_VGG16_TensorFlow 中的代码进行操作。

对于 TensorFlow 中 VGG16 的所有示例，首先从 http://download.tensorflow.org/models/vgg_16_2016_08_28.tar.gz 下载检查点文件，并使用如下代码初始化变量：

```
model_name='vgg_16'
model_url='http://download.tensorflow.org/models/'
model_files=['vgg_16_2016_08_28.tar.gz']
model_home=os.path.join(models_root,model_name)

dsu.download_dataset(source_url=model_url,
    source_files=model_files,
    dest_dir = model_home,
    force=False,
    extract=True)
```

再定义一些通用的变量并导入相应的包：

```
from tensorflow.contrib import slim
from tensorflow.contrib.slim.nets import vgg
image_height=vgg.vgg_16.default_image_size
image_width=vgg.vgg_16.default_image_size
```

12.4.1 使用 TensorFlow 中预先训练的 VGG16 进行图像分类

现在不是进行重新训练,而是尝试预测测试图像的类别(见表 12-3)。首先,清除默认图并定义图像的占位符:

```
tf.reset_default_graph()
x_p = tf.placeholder(shape=(None,image_height, image_width,3),
                     dtype=tf.float32,name='x_p')
```

占位符 x_p 的形状是(?,224,224,3)。接下来,加载 vgg16 模型:

```
with slim.arg_scope(vgg.vgg_arg_scope()):
    logits,_ = vgg.vgg_16(x_p,num_classes=inet.n_classes,
                         is_training=False)
```

添加 softmax 层生成类别的概率:

```
probabilities = tf.nn.softmax(logits)
```

定义初始化函数来恢复变量,例如检查点文件中的权重和偏差。

```
init = slim.assign_from_checkpoint_fn(
    os.path.join(model_home, '{}.ckpt'.format(model_name))
```

在 TensorFlow 会话中,初始化变量并运行概率张量获取每幅图像的概率:

```
with tf.Session() as tfs:
    init(tfs)
    probs = tfs.run([probabilities],feed_dict={x_p:images_test})
    probs=probs[0]
```

显示得到的类别:

```
disp(images_test,id2label=inet.id2label,probs=probs,scale=True)
```

表 12-3

输入图像	输出概率
	为[斑马]的概率是 99.15% 为[老虎,猫]的概率是 0.37% 为[老虎,豹,东北虎]的概率是 0.33% 为[鹅]的概率是 0.04% 为[虎斑猫,虎斑猫]的概率是 0.02%

（续）

输入图像	输出概率
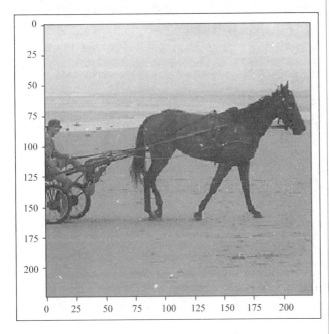	为[马车]的概率是 99.50% 为[犁]的概率是 0.37% 为[阿拉伯骆驼,单峰骆驼,骆驼]的概率是 0.06% 为[栗色马]的概率是 0.05% 为[桶]的概率是 0.01%
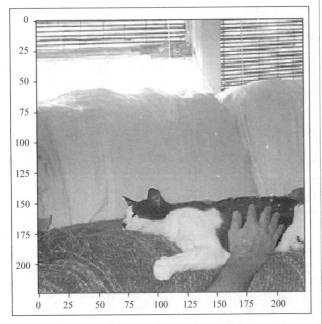	为[羊毛衫,卡迪根威尔士柯基犬]的概率是 9.32% 为[蝴蝶犬]的概率是 11.78% 为[设得兰牧羊犬]的概率是 9.01% 为[暹罗猫]的概率是 7.09% 为[彭布罗克威尔士柯基犬]的概率是 6.27%

(续)

输入图像	输出概率
	为［山雀］的概率是 97.09% 为［河鸟］的概率是 2.52%的 为［灯芯草雀，雪鸟］的概率是 0.23% 为［蜂鸟］的概率是 0.09% 为［夜莺］的概率是 0.04%
	为［惠比特犬］的概率是 24.98% 为［狮子］的概率是 16.48% 为［萨卢基狗，瞪羚猎犬］的概率是 5.54% 为［棕熊，布伦熊］的概率是 4.99% 为［丝毛猎狐犬］的概率是 4.11%

（续）

输入图像	输出概率
	为［棕熊，布伦熊］的概率是 98.56% 为［美洲黑熊，黑熊］的概率是 1.40% 为［懒熊］的概率是 0.03% 为［袋熊］的概率是 0.00% 为［河狸］的概率是 0.00%
	为［豹，金钱豹］的概率是 20.84% 为［猎豹，印度豹］的概率是 12.81% 为［条纹壁虎］的概率是 12.26% 为［美洲虎，黑豹，美洲豹］的概率是 10.28% 为［瞪羚］的概率是 5.30%

输入图像	输出概率
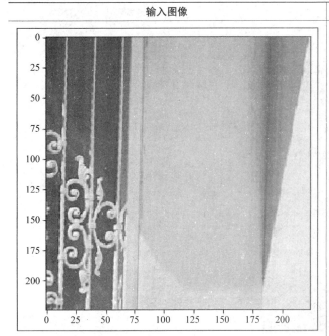	为［浴帘］概率是 8.09% 为［粘合剂，环状粘合剂］的概率是 3.59% 为［手风琴，键盘式手风琴，六角形手风琴］概率是 3.32% 为［散热器］的概率是 3.12% 为［长袍］的概率是 1.81%

预训练模型从未见过数据集中的图像，并且不了解数据集中的类别，但它却能正确识别了斑马、马车、鸟和熊，只是不能认出长颈鹿，因为模型以前从未见过长颈鹿。要在数据集上重新训练这个模型，只需要更少的工作量和较小的数据集（包含 800 幅图像）。但在重新训练模型之前，先需要在 TensorFlow 中进行相同的图像预处理。

12.5 将 TensorFlow 中的图像预处理用于预先训练的 VGG16

在 TensorFlow 中为预处理步骤定义一个函数，具体的代码如下：

```
def tf_preprocess(filelist):
    images=[]
    for filename in filelist:
        image_string = tf.read_file(filename)
        image_decoded = tf.image.decode_jpeg(image_string, channels=3)
        image_float = tf.cast(image_decoded, tf.float32)
        resize_fn = tf.image.resize_image_with_crop_or_pad
        image_resized = resize_fn(image_float, image_height, image_width)
        means = tf.reshape(tf.constant([123.68, 116.78, 103.94]),
                           [1, 1, 3])
        image = image_resized - means
        images.append(image)
    images = tf.stack(images)
    return images
```

这里创建了 images 变量而不是占位符：

```
        images=tf_preprocess([x for x in x_test])
```

按照与以前相同的过程定义 VGG16 模型，保存变量，然后进行预测：

```
    with slim.arg_scope(vgg.vgg_arg_scope()):
        logits,_ = vgg.vgg_16(images,
                              num_classes=inet.n_classes,
                              is_training=False
                              )
    probabilities = tf.nn.softmax(logits)

    init = slim.assign_from_checkpoint_fn(
            os.path.join(model_home, '{}.ckpt'.format(model_name)),
            slim.get_variables_to_restore())
```

这里会获得与前面相同的预测概率。给出上述操作只是想证明预处理也可以在 TensorFlow 中完成。但是，TensorFlow 中的预处理仅限于 TensorFlow 提供的函数，这与该框架有紧密的联系。

建议读者将预处理过程与 TensorFlow 的模型训练预测的代码分开。保持独立可以使其具有模块化并具有其他优势，例如：为了在多个模型中重复使用，可以保存数据。

12.5.1 使用 TensorFlow 中重新训练的 VGG16 进行图像分类

现在，用基于 COCO 的动物数据集重新训练 VGG16 模型（见表 12-4）。下面定义了三个占位符：

- is_training 占位符指定是否将模型用于训练或预测；
- x_p 是输入的占位符形状（None, image_height, image_width, 3）；
- y_p 是输出的占位符形状（None,1）。

```
        is_training = tf.placeholder(tf.bool,name='is_training')
        x_p = tf.placeholder(shape=(None,image_height, image_width,3),
                             dtype=tf.float32,name='x_p')
        y_p = tf.placeholder(shape=(None,1),dtype=tf.int32,name='y_p')
```

正如前面介绍原理部分时所解释的那样，除了最后一层（称为 vgg / fc8 层）之外，将从检查点文件恢复网络层：

```
with slim.arg_scope(vgg.vgg_arg_scope()):
    logits, _ = vgg.vgg_16(x_p,num_classes=coco.n_classes,
                           is_training=is_training)

probabilities = tf.nn.softmax(logits)
# 恢复，除了最后一层fc8
fc7_variables=tf.contrib.framework.get_variables_to_restore(exclude=['vgg_16/fc8'])
fc7_init = tf.contrib.framework.assign_from_checkpoint_fn(
    os.path.join(model_home, '{}.ckpt'.format(model_name)),
    fc7_variables)
```

接下来，定义最后一层的变量，这些变量需要初始化：

```
# fc8层
fc8_variables = tf.contrib.framework.get_variables('vgg_16/fc8')
fc8_init = tf.variables_initializer(fc8_variables)
```

像前面章节那样使用 tf.losses.sparse_softmax_cross_entropy( ) 作为损失函数。

```
tf.losses.sparse_softmax_cross_entropy(labels=y_p, logits=logits)
loss = tf.losses.get_total_loss()
```

对最后一层运行几个迭代周期进行训练，然后对整个网络的几层进行训练。因此，定义两个单独的优化器和训练操作。

```
learning_rate = 0.001
fc8_optimizer = tf.train.GradientDescentOptimizer(learning_rate)
fc8_train_op = fc8_optimizer.minimize(loss, var_list=fc8_variables)

full_optimizer = tf.train.GradientDescentOptimizer(learning_rate)
full_train_op = full_optimizer.minimize(loss)
```

对两个优化器函数使用相同的学习率，若想进一步调整超参数，则可以定义单独的学习率。

像前面一样定义精度函数：

```
y_pred = tf.to_int32(tf.argmax(logits, 1))
n_correct_pred = tf.equal(y_pred, y_p)
accuracy = tf.reduce_mean(tf.cast(n_correct_pred, tf.float32))
```

最后，通过 10 个周期训练最后一层网络，然后对整个网络进行批大小为 32、周期为 10 的训练。也使用相同的会话预测类别：

```
fc8_epochs = 10
full_epochs = 10
coco.y_onehot = False
coco.batch_size = 32
coco.batch_shuffle = True

total_images = len(x_train_files)
n_batches = total_images // coco.batch_size

with tf.Session() as tfs:
    fc7_init(tfs)
    tfs.run(fc8_init)
    for epoch in range(fc8_epochs):
        print('Starting fc8 epoch ',epoch)
        coco.reset_index()
        epoch_accuracy=0
        for batch in range(n_batches):
            x_batch, y_batch = coco.next_batch()
            images=np.array([coco.preprocess_for_vgg(x) \
                    for x in x_batch])
            feed_dict={x_p:images,y_p:y_batch,is_training:True}
            tfs.run(fc8_train_op, feed_dict = feed_dict)
```

```
            feed_dict={x_p:images,y_p:y_batch,is_training:False}
            batch_accuracy = tfs.run(accuracy,feed_dict=feed_dict)
            epoch_accuracy += batch_accuracy
        except Exception as ex:
    epoch_accuracy /= n_batches
    print('Train accuracy in epoch {}:{}'
          .format(epoch,epoch_accuracy))
for epoch in range(full_epochs):
    print('Starting full epoch ',epoch)
    coco.reset_index()
    epoch_accuracy=0
    for batch in range(n_batches):
        x_batch, y_batch = coco.next_batch()
        images=np.array([coco.preprocess_for_vgg(x) \
                for x in x_batch])
        feed_dict={x_p:images,y_p:y_batch,is_training:True}
        tfs.run(full_train_op, feed_dict = feed_dict )
        feed_dict={x_p:images,y_p:y_batch,is_training:False}
        batch_accuracy = tfs.run(accuracy,feed_dict=feed_dict)
        epoch_accuracy += batch_accuracy
    epoch_accuracy /= n_batches
    print('Train accuracy in epoch {}:{}'
          .format(epoch,epoch_accuracy))
# 现在进行预测
feed_dict={x_p:images_test,is_training: False}
probs = tfs.run([probabilities],feed_dict=feed_dict)
probs=probs[0]
```

输出预测结果：

```
disp(images_test,id2label=coco.id2label,probs=probs,scale=True)
```

表 12-4

输入图像	输出概率
	为［斑马］的概率是 100.00%

（续）

输入图像	输出概率
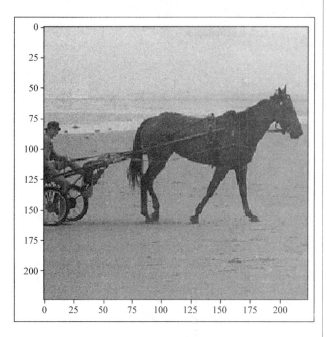	为［马］的概率是 100.00%
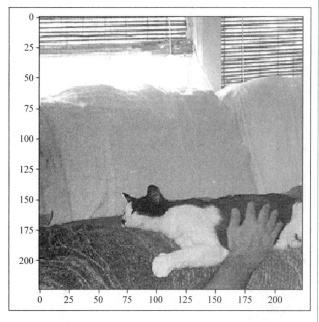	为［猫］的概率是 98.88%

（续）

输入图像	输出概率
	为［鸟］的概率是 100.00%
	为［熊］的概率是 68.88% 为［绵羊］的概率是 31.06% 为［狗］的概率是 0.02% 为［鸟］的概率是 0.02% 为［马］的概率是 0.01%

（续）

输入图像	输出概率
	为［熊］的概率是 100.00% 为［狗］的概率是 0.00% 为［鸟］的概率是 0.00% 为［绵羊］的概率是 0.00% 为［猫］的概率是 0.00%
	为［长颈鹿］的概率是 100.00%

(续)

输入图像	输出概率
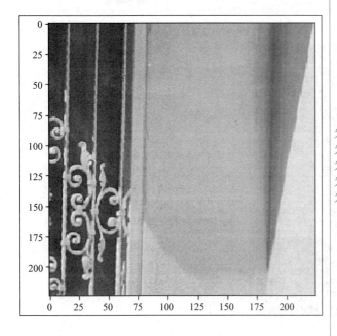	为［猫］的概率是 61.36% 为［狗］的概率是 16.70% 为［鸟］的概率是 7.46% 为［熊］的概率是 5.34% 为［长颈鹿］的概率是 3.65%

模型正确识别了猫和长颈鹿，并将识别的概率提高到 100%。但仍然犯了一些错误，因为最后的照片被归类为猫，这实际上是裁剪后的含噪声图片。可以根据需求进一步改进这些结果。

12.6　Keras 中的 VGG16

可以根据 Jupyter 笔记本 ch-12a_VGG16_Keras 中的代码进行操作。

现在用 Keras 进行同样的分类和重新训练，可以看到在 Keras 中，使用 VGG16 预训练模型只需要较少的代码。

12.6.1　使用 Keras 中预先训练的 VGG16 进行图像分类

可用下面这行代码加载模型：

```
from keras.applications import VGG16
model=VGG16(weights='imagenet')
```

可以使用这个模型预测类别的概率：

```
probs = model.predict(images_test)
```

表 12-5 是分类结果。

表 12-5

输入图像	输出概率
	为［斑马］的概率是 99.41% 为［老虎，猫］的概率是 0.19% 为［鹅］的概率是 0.13% 为［老虎，孟加拉虎］的概率是 0.09% 为［蘑菇］的概率是 0.02%
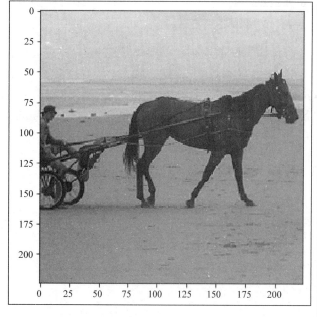	为［马车］的概率是 87.50% 为［阿拉伯骆驼，单峰骆驼］的概率是 5.58% 为［犁］的概率是 4.72% 为［狗拉雪橇］的概率是 1.03% 为［破车］的概率是 0.31%

（续）

输入图像	输出概率
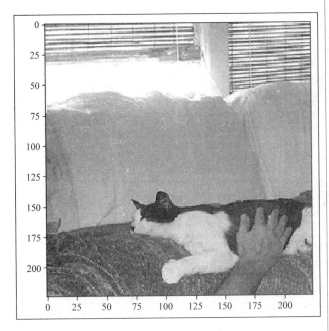	为［暹罗猫］的概率是 34.96% 为［小猎犬］的概率是 12.71% 为［波士顿公牛，波士顿狗］的概率是 10.15% 为［意大利格雷猎狗］的概率是 6.53% 为［卡迪根威尔士柯基犬］的概率是 6.01%
	为［灯芯草雀，雪鸟］的概率是 56.41% 为［山雀］的概率是 38.08% 为［夜莺］的概率是 1.93% 为［蜂鸟］的概率是 1.35% 为［家朱雀，红雀］的概率是 1.09%

（续）

输入图像	输出概率
	为［棕熊，布伦熊］的概率是 54.19% 为［狮子］的概率是 28.07% 为［诺里奇梗犬］的概率是 0.87% 为［湖地犬］的概率是 0.82% 为［野猪］的概率是 0.73%
	为［棕熊，布伦熊］的概率是 88.64% 为［美洲黑熊，黑熊］的概率是 7.22% 为［懒熊］的概率是 4.13% 为［獾］的概率是 0.00% 为［袋熊］的概率是 0.00%

(续)

输入图像	输出概率
	为 [美洲虎，黑豹，美洲豹] 的概率是 38.70% 为 [豹，金钱豹] 的概率是 33.78% 为 [猎豹，印度豹] 的概率是 14.22% 为 [条纹壁虎] 的概率是 6.15% 为 [雪豹] 的概率是 1.53%
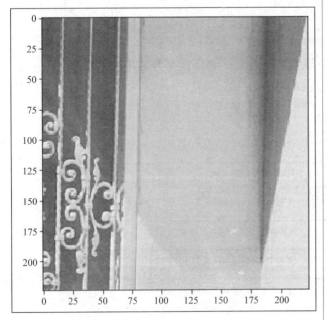	为 [浴帘] 的概率是 12.54% 为 [粘合剂，环状粘合剂] 的概率是 2.82% 为 [卫生纸] 的概率是 2.82% 为 [手风琴，键盘式手风琴，六角形手风琴] 的概率是 2.12% 为 [浴巾] 的概率是 2.05%

该模型无法识别绵羊、长颈鹿而且最后一张含噪声图像（由狗的图像裁剪而成）。现在用数据集重新训练 Keras 中的模型。

12.6.2 使用 Keras 中重新训练的 VGG16 进行图像分类

为了微调分类任务，本小节将使用 COCO 图像数据集重新训练模型。删除 Keras 模型中的最后一层，并使用 softmax 激活函数作为全连接层实现 8 个分类器。为了演示固定前几层的情形，可将前 15 层的可训练属性设置为 False。

1）首先将 include_top 设置为 False，导入 VGG16 模型，这不会使用顶层变量：

```
# 加载vgg模型
from keras.applications import VGG16
base_model=VGG16(weights='imagenet',include_top=False,
input_shape=(224,224,3))
```

上面的代码还需要指定 input_shape，否则 Keras 会在以后抛出异常。

2）在导入的 VGG 模型之上构建分类器模型：

```
top_model = Sequential()
top_model.add(Flatten(input_shape=base_model.output_shape[1:]))
top_model.add(Dense(256, activation='relu'))
top_model.add(Dropout(0.5))
top_model.add(Dense(coco.n_classes, activation='softmax'))
```

3）接下来，在 VGG 基础上添加模型：

```
model=Model(inputs=base_model.input,
outputs=top_model(base_model.output))
```

4）固定前 15 层：

```
for layer in model.layers[:15]:
    layer.trainable = False
```

5）随机挑选要固定的 15 层，这个数字可以改变。编译模型并输出模型概要：

```
model.compile(loss='categorical_crossentropy',
        optimizer=optimizers.SGD(lr=1e-4, momentum=0.9),
        metrics=['accuracy'])
model.summary()
```

Layer (type)	Output Shape	Param #
input_1 (InputLayer)	(None, 224, 224, 3)	0
block1_conv1 (Conv2D)	(None, 224, 224, 64)	1792
block1_conv2 (Conv2D)	(None, 224, 224, 64)	36928
block1_pool (MaxPooling2D)	(None, 112, 112, 64)	0
block2_conv1 (Conv2D)	(None, 112, 112, 128)	73856

block2_conv2 (Conv2D)	(None, 112, 112, 128)	147584
block2_pool (MaxPooling2D)	(None, 56, 56, 128)	0
block3_conv1 (Conv2D)	(None, 56, 56, 256)	295168
block3_conv2 (Conv2D)	(None, 56, 56, 256)	590080
block3_conv3 (Conv2D)	(None, 56, 56, 256)	590080
block3_pool (MaxPooling2D)	(None, 28, 28, 256)	0
block4_conv1 (Conv2D)	(None, 28, 28, 512)	1180160
block4_conv2 (Conv2D)	(None, 28, 28, 512)	2359808
block4_conv3 (Conv2D)	(None, 28, 28, 512)	2359808
block4_pool (MaxPooling2D)	(None, 14, 14, 512)	0
block5_conv1 (Conv2D)	(None, 14, 14, 512)	2359808
block5_conv2 (Conv2D)	(None, 14, 14, 512)	2359808
block5_conv3 (Conv2D)	(None, 14, 14, 512)	2359808
block5_pool (MaxPooling2D)	(None, 7, 7, 512)	0
sequential_1 (Sequential)	(None, 8)	6424840

```
Total params: 21,139,528
Trainable params: 13,504,264
Non-trainable params: 7,635,264
```

可以看到将近40%的参数被固定并且是不可训练的。

6）接下来，用Keras对整个模型进行训练，周期为20，批大小为32。

```
from keras.utils import np_utils

batch_size=32
n_epochs=20

total_images = len(x_train_files)
n_batches = total_images // batch_size
for epoch in range(n_epochs):
    print('Starting epoch ',epoch)
    coco.reset_index_in_epoch()
    for batch in range(n_batches):
        try:
            x_batch, y_batch = coco.next_batch(batch_size=batch_size)
            images=np.array([coco.preprocess_image(x) for x in x_batch])
```

```
                y_onehot = np_utils.to_categorical(y_batch,
                        num_classes=coco.n_classes)
                model.fit(x=images,y=y_onehot,verbose=0)
        except Exception as ex:
            print('error in epoch {} batch {}'.format(epoch,batch))
            print(ex)
```

7）使用重新训练的模型对图像进行分类：

```
        probs = model.predict(images_test)
```

表 12-6 是分类结果。

表 12-6

输入图像	输出概率
	为［斑马］的概率是 100.00% 为［狗］的概率是 0.00% 为［马］的概率是 0.00% 为［长颈鹿］的概率是 0.00% 为［熊］的概率是 0.00%
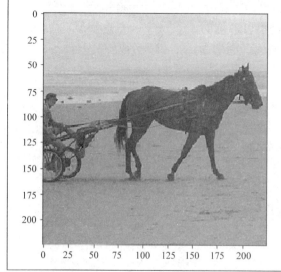	为［马］的概率是 96.11% 为［猫］的概率是 1.85% 为［鸟］的概率是 0.77% 为［长颈鹿］的概率是 0.43% 为［绵羊］的概率是 0.40%

（续）

输入图像	输出概率
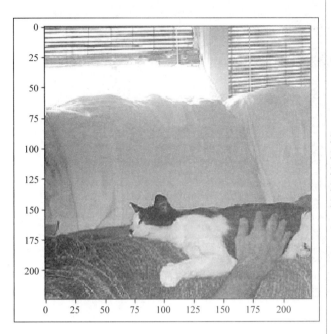	为［狗］的概率是 99.75% 为［猫］的概率是 0.22% 为［马］的概率是 0.03% 为［熊］的概率是 0.00% 为［斑马］的概率是 0.00%
	为［鸟］的概率是 99.88% 为［马］的概率是 0.11% 为［长颈鹿］的概率是 0.00% 为［熊］的概率是 0.00% 为［猫］的概率是 0.00%

（续）

输入图像	输出概率
	为［熊］的概率是 65.28% 为［绵羊］的概率是 27.09% 为［鸟］的概率是 4.34% 为［长颈鹿］的概率是 1.71% 为［狗］的概率是 0.63%
	为［熊］的概率是 100.00% 为［绵羊］的概率是 0.00% 为［狗］的概率是 0.00% 为［猫］的概率是 0.00% 为［长颈鹿］的概率是 0.00%

（续）

输入图像	输出概率
	为［长颈鹿］的概率是 100.00% 为［鸟］的概率是 0.00% 为［熊］的概率是 0.00% 为［绵羊］的概率是 0.00% 为［斑马］的概率是 0.00%
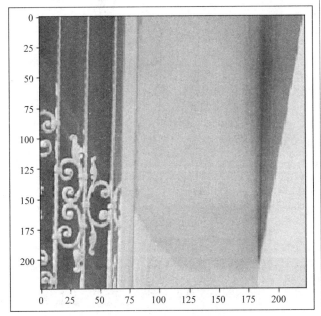	为［猫］的概率是 81.05% 为［狗］的概率是 15.68% 为［鸟］的概率是 1.64% 为［马］的概率是 0.90% 为［熊］的概率是 0.43%

除最后一个含噪声图像外,所有类别都已正确识别。通过适当调整超参数进行改进。

到目前为止,已经看到了使用预训练模型进行分类并对预训练模型进行微调的示例。接下来,将介绍使用 Inception v3 模型进行分类的示例。

12.7 TensorFlow 中的 Inception v3

可以按照 Jupter 笔记本 ch-12c_InceptionV3_TensorFlow 中代码进行操作。

TensorFlow 的 Inception v3 在 1 001 个标签上训练,而不是 1 000 个。此外,会用不同的方法预处理训练图像。这里会展示上一节的预处理代码。然后会深入介绍如何通过 TensorFlow 恢复 Inception v3 模型。

下载 Inception v3 的检查点文件:

```
# 加载Inception v3模型
model_name='inception_v3'
model_url='http://download.tensorflow.org/models/'
model_files=['inception_v3_2016_08_28.tar.gz']
model_home=os.path.join(models_root,model_name)

dsu.download_dataset(source_url=model_url,
    source_files=model_files,
    dest_dir = model_home,
    force=False,
    extract=True)
```

定义 Inception 模块和变量的常见导入:

```
### 定义常见导入和变量
from tensorflow.contrib.slim.nets import inception
image_height=inception.inception_v3.default_image_size
image_width=inception.inception_v3.default_image_size
```

12.7.1 使用 TensorFlow 中 Inception v3 进行图像分类

图像分类的原理与上一节使用 VGG 16 模型一样。Inception v3 模型的完整代码如下:

```
x_p = tf.placeholder(shape=(None,
                            image_height,
                            image_width,
                            3
                           ),
                     dtype=tf.float32,
                     name='x_p')
with slim.arg_scope(inception.inception_v3_arg_scope()):
    logits,_ = inception.inception_v3(x_p,
                                      num_classes=inet.n_classes,
```

```
                            is_training=False
                            )
probabilities = tf.nn.softmax(logits)

init = slim.assign_from_checkpoint_fn(
        os.path.join(model_home, '{}.ckpt'.format(model_name)),
        slim.get_variables_to_restore())
with tf.Session() as tfs:
    init(tfs)
    probs = tfs.run([probabilities],feed_dict={x_p:images_test})
    probs=probs[0]
```

下面看看模型如何处理测试图像，如表 12-7 所示。

表 12-7

输入图像	输出概率
	为［斑马］的概率是 95.15% 为［鸵鸟］的概率是 0.07% 为［麋羚］的概率是 0.07% 为［短袜］的概率是 0.03% 为［疣猪］的概率是 0.03%
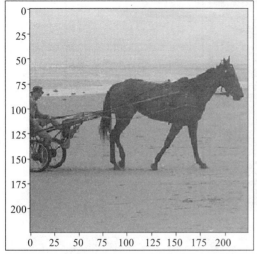	为［马车］的概率是 93.09% 为［犁］的概率是 0.47% 为［牛车］的概率是 0.07% 为［海滨，海岸］的概率是 0.07% 为［军装］的概率是 0.06%

（续）

输入图像	输出概率
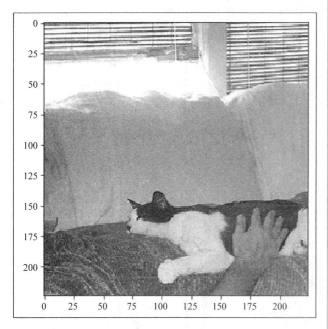	为［羊毛衫，卡迪根威尔士柯基犬］的概率是18.94% 为［彭布罗克，彭布罗克威尔士柯基犬］的概率是8.19% 为［工作室沙发，沙发床］的概率是7.86% 为［英国斯普林格，英国史宾格犬］的概率是5.36% 为［边境牧羊犬］的概率是4.16%
	为［河乌］的概率是27.18% 为［雪鸟］的概率是24.38% 为［山雀］的概率是6.91% 为［喜鹊］的概率是0.99% 为［燕雀］的概率是0.73%

输入图像	输出概率
	为［猪，雏鸽，野猪］的概率是 93.00% 为［野猪］的概率是 2.23% 为［公羊］的概率是 0.65% 为［公牛］的概率是 0.43% 为［土拨鼠］的概率是 0.23%
	为［棕熊，布伦熊］的概率是 84.27% 为［美洲黑熊，黑熊］的概率是 1.57% 为［懒熊］的概率是 1.34% 为［小熊猫］的概率是 0.13% 为［北极熊］的概率是 0.12%

(续)

输入图像	输出概率
	为[蜂窝]的概率是 20.20% 为[瞪羚]的概率是 6.52% 为[栗色马]的概率是 5.14% 为[黑斑羚]的概率是 3.72% 为[萨卢基犬，瞪羚猎犬]的概率是 2.44%
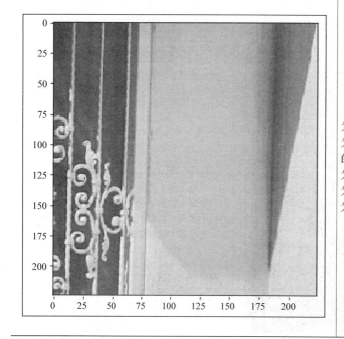	为[竖琴]的概率是 41.17% 为[手风琴，键盘式手风琴，六角形手风琴]的概率是 13.64% 为[遮光帘]的概率是 2.97% 为[链子]的概率是概率 1.59% 为[付费电话，付费电台]的概率是 1.55%

虽然该模型在与 VGG 模型几乎相同的地方失败了，但也不算太糟糕。现在用 COCO 动物图像和标签重新训练这个模型。

12.7.2 使用 TensorFlow 中重新训练的 Inception v3 进行图像分类

Inception v3 的重新训练与 VGG16 不同，因为这里使用 softmax 激活层作为输出，并使用 tf.losses.softmax_cross_entropy( ) 作为损失函数。

1）首先，定义占位符：

```
is_training = tf.placeholder(tf.bool,name='is_training')
x_p = tf.placeholder(shape=(None,
                            image_height,
                            image_width,
                            3
                            ),
                    dtype=tf.float32,
                    name='x_p')
y_p = tf.placeholder(shape=(None,coco.n_classes),
                    dtype=tf.int32,
                    name='y_p')
```

2）加载模型

```
with slim.arg_scope(inception.inception_v3_arg_scope()):
    logits,_ = inception.inception_v3(x_p,
                                      num_classes=coco.n_classes,
                                      is_training=True
                                      )
probabilities = tf.nn.softmax(logits)
```

3）接下来，定义函数恢复除最后一层之外的变量：

```
with slim.arg_scope(inception.inception_v3_arg_scope()):
    logits,_ = inception.inception_v3(x_p,
                                      num_classes=coco.n_classes,
                                      is_training=True
                                      )
probabilities = tf.nn.softmax(logits)
# 恢复，除最后一层
checkpoint_exclude_scopes=["InceptionV3/Logits",
                           "InceptionV3/AuxLogits"]
exclusions = [scope.strip() for scope in checkpoint_exclude_scopes]

variables_to_restore = []
for var in slim.get_model_variables():
    excluded = False
    for exclusion in exclusions:
        if var.op.name.startswith(exclusion):
            excluded = True
            break
    if not excluded:
        variables_to_restore.append(var)

init_fn = slim.assign_from_checkpoint_fn(
    os.path.join(model_home, '{}.ckpt'.format(model_name)),
    variables_to_restore)
```

4）定义损失函数、优化器和训练过程：

```
tf.losses.softmax_cross_entropy(onehot_labels=y_p, logits=logits)
loss = tf.losses.get_total_loss()
learning_rate = 0.001
optimizer = tf.train.GradientDescentOptimizer(learning_rate)
train_op = optimizer.minimize(loss)
```

5）训练模型并在训练完成之后，在同一会话中进行预测：

```
n_epochs=10
coco.y_onehot = True
coco.batch_size = 32
coco.batch_shuffle = True
total_images = len(x_train_files)
n_batches = total_images // coco.batch_size

with tf.Session() as tfs:
    tfs.run(tf.global_variables_initializer())
    init_fn(tfs)
    for epoch in range(n_epochs):
        print('Starting epoch ',epoch)
        coco.reset_index()
        epoch_accuracy=0
        epoch_loss=0
        for batch in range(n_batches):
            x_batch, y_batch = coco.next_batch()
            images=np.array([coco.preprocess_for_inception(x) \
                for x in x_batch])
            feed_dict={x_p:images,y_p:y_batch,is_training:True}
            batch_loss,_ = tfs.run([loss,train_op],
                            feed_dict = feed_dict)
            epoch_loss += batch_loss
        epoch_loss /= n_batches
        print('Train loss in epoch {}:{}'
            .format(epoch,epoch_loss))
    #  现在进行预测
    feed_dict={x_p:images_test,is_training: False}
    probs = tfs.run([probabilities],feed_dict=feed_dict)
    probs=probs[0]
```

可以看到每次迭代的损失函数值都在减少：

```
INFO:tensorflow:Restoring parameters from
/home/armando/models/inception_v3/inception_v3.ckpt
Starting epoch  0
Train loss in epoch 0:2.7896385192871094
Starting epoch  1
Train loss in epoch 1:1.6651896286010741
Starting epoch  2
Train loss in epoch 2:1.2332031989097596
Starting epoch  3
Train loss in epoch 3:0.9912329530715942
Starting epoch  4
    Train loss in epoch 4:0.8110128355026245
```

```
Starting epoch   5
Train loss in epoch 5:0.7177265572547913
Starting epoch   6
Train loss in epoch 6:0.6175705575942994
Starting epoch   7
Train loss in epoch 7:0.5542363750934601
Starting epoch   8
Train loss in epoch 8:0.523461252450943
Starting epoch   9
Train loss in epoch 9:0.4923107647895813
```

这次结果正确识别了绵羊，但错误地将猫图片识别为狗，如表 12-8 所示。

表 12-8

输入图像	输出概率
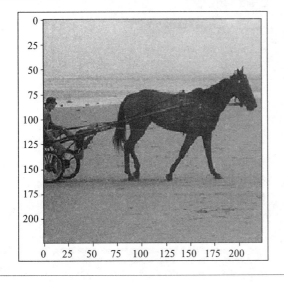	为［斑马］的概率是 98.84% 为［长颈鹿］的概率是 0.84% 为［绵羊］的概率是 0.11% 为［猫］的概率是 0.07% 为［狗］的概率是 0.06%
	为［马］的概率是 95.77% 为［狗］的概率是 1.34% 为［斑马］的概率是 0.89% 为［鸟］的概率是 0.68% 为［绵羊］的概率是 0.61%

（续）

输入图像	输出概率
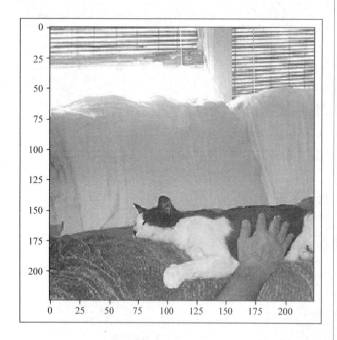	为［狗］的概率是 94.83% 为［猫］的概率是 4.53% 为［绵羊］的概率是 0.56% 为［熊］的概率是 0.04% 为［斑马］的概率是 0.02%
	为［鸟］的概率是 42.80% 为［猫］的概率是 25.64% 为［熊］的概率是 15.56% 为［长颈鹿］的概率是 8.77% 为［绵羊］的概率是 3.39%

（续）

输入图像	输出概率
	为［绵羊］的概率是 72.58% 为［熊］的概率是 8.40% 为［长颈鹿］的概率是 7.64% 为［马］的概率是 4.02% 为［鸟］的概率是 3.65%
	为［熊］的概率是 98.03% 为［猫］的概率是 0.74% 为［绵羊］的概率是 0.54% 为［鸟］的概率是 0.28% 为［马］的概率是 0.17%

（续）

输入图像	输出概率
	为［长颈鹿］的概率是 96.43% 为［鸟］的概率是 1.78% 为［绵羊］的概率是 1.10% 为［斑马］的概率是 0.32% 为［熊］的概率是 0.14%
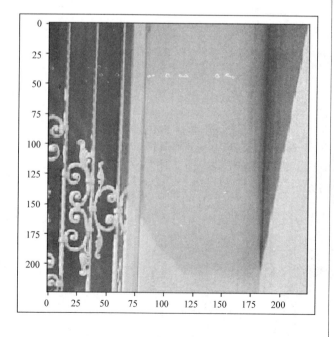	为［马］的概率是 34.43% 为［狗］的概率是 23.53% 为［斑马］的概率是 16.03% 为［猫］的概率是的 9.76% 为［长颈鹿］的概率是 9.02%

12.8 总结

迁移学习是一项伟大的发现，它允许我们通过将较大数据集中训练好的模型应用于不同数据集上的模型训练，从而节省时间。当数据集很小时，迁移学习也有助于热启动训练过程。本章学习了如何使用预先训练的模型（如 VGG16 和 Inception v3）针对不同数据集训练用于图像分类的模型；还学习了如何使用 TensorFlow 和 Keras 中的示例重新训练预训练模型，以及如何为不同模型进行预处理图像。

本章还介绍了几种在 ImageNet 数据集上训练出的模型。读者可以尝试查找在不同数据集（例如视频数据集、语音数据集或文本 / NLP 数据集）上训练的其他模型。读者也可以尝试使用这些模型重新训练，并可用于在自己的数据集上训练深度学习模型。

第 13 章
深度强化学习

强化学习是一种学习形式,即软件代理(agent)观察环境并采取行动,以最大限度地从环境中得到奖励,如图 13-1 所示。

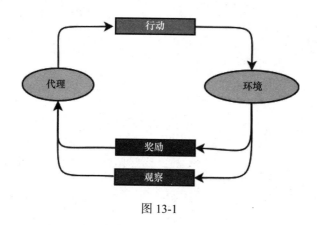

图 13-1

这个观点在下面这些现实生活中会碰到:
- 股票交易代理通过观察交易信息、新闻、报告和其他形式信息,然后实施买入或卖出的交易行为,以短期或长期利润的形式最大化奖励。
- 保险代理软件分析客户的信息,然后计算保险费金额,以便最大化利润并最大限度地降低风险。
- 机器人观察环境后会采取行动,例如步行、跑步或拾取物体,以达到目标作为最大化的回报。

强化学习已经成功应用于多个领域,例如广告优化、股票市场交易、自动驾驶车辆、机器人和游戏等。

强化学习与监督学习不同,它预先没有标签可以用于调整模型参数。强化模型通过在运行中获得的奖励进行学习。虽然可立即获得短期奖励,但要获得长期奖励需要经过几个步骤。这种现象也被称为**延迟反馈**。

强化学习也与无监督学习不同,因为在无监督学习中没有可用的标签,而在强化学习中,会以奖励的形式作为反馈。

本章将通过以下主题介绍强化学习及其在 TensorFlow 和 Keras 中的实现:

- OpenAI Gym [1] 101。
- 将简单的策略应用于 cartpole 游戏。
- 强化学习 101：
  - Q 函数（在模型无效时学习优化）；
  - 强化学习算法的探索与开发；
  - V 函数（在模型可用时学习优化）；
  - 强化学习技巧。
- 强化学习的朴素神经网络策略。
- 实施 Q-Learning：
  - Q-Learning 的初始化和离散化；
  - 基于 Q 表的 Q-Learning；
  - 使用 Q 网络或深度 Q 网络（DQN）进行 Q-Learning

本章将在 OpenAI Gym 中演示示例，所以需要首先介绍 OpenAI Gym。

13.1　OpenAI Gym 101

OpenAI Gym 是一个基于 Python 的工具包，用于研究和开发强化学习算法。在撰写本书时，OpenAI Gym 提供了 700 多个开源环境。使用 OpenAI，还可以创建自己的环境。OpenAI Gym 最大的优势是它可以提供统一的界面来处理这些环境，用户可专注于强化学习算法，其运行模拟可交给 OpenAI Gym。

介绍 OpenAI Gym 的研究论文可通过以下链接获得：http://arxiv.org/abs/1606.01540。

可以使用以下命令安装 OpenAI Gym：

```
pip3 install gym
```

如果上述命令不起作用，则可以通过以下链接找到有关安装的帮助信息：https://github.com/openai/gym#installation。

1）在 OpenAI Gym 中输出可用的环境的数量：

可以按照 Jupyter 笔记本 ch-13a_Reinforcement_Learning_NN 中的代码进行操作。

---

[1] OpenAI Gym 是一款用于研发和比较各种强化学习算法的工具包。——译者注

```
all_env = list(gym.envs.registry.all())
print('Total Environments in Gym version {} : {}'
    .format(gym.__version__,len(all_env)))

Total Environments in Gym version 0.9.4 : 777
```

2）得到所有环境的列表：

```
for e in list(all_env):
    print(e)
```

部分输出结果如下：

```
EnvSpec(Carnival-ramNoFrameskip-v0)
EnvSpec(EnduroDeterministic-v0)
EnvSpec(FrostbiteNoFrameskip-v4)
EnvSpec(Taxi-v2)
EnvSpec(Pooyan-ram-v0)
EnvSpec(Solaris-ram-v4)
EnvSpec(Breakout-ramDeterministic-v0)
EnvSpec(Kangaroo-ram-v4)
EnvSpec(StarGunner-ram-v4)
EnvSpec(Enduro-ramNoFrameskip-v4)
EnvSpec(DemonAttack-ramDeterministic-v0)
EnvSpec(TimePilot-ramNoFrameskip-v0)
EnvSpec(Amidar-v4)
```

由 env 对象表示的每个环境都有一个标准化的接口，例如：
- env.make(<game-id-string>) 函数通过传递 id 字符串创建一个 env 对象。
- 每个 env 对象包含以下主要函数：
  - step( ) 函数将 action 对象作为参数并返回四个对象：
    - observation：由环境实现的对象，代表对环境的观察。
    - reward：一个有符号的浮点值，表示前一个操作的收益（或损失）。
    - done：取布尔值。表示方案是否已完成。
    - info：表示诊断信息的 Python 字典对象。
  - render( ) 函数创建环境的可视化表示。
  - reset( ) 函数将环境重置为原始状态。
- 每个 env 对象都带有明确定义的动作和观察，由 action_space 和 observation_space 表示。

CartPole 是 gym 包中最受欢迎的强化学习游戏之一。在这个游戏中，连接到推车的杆子必须平衡，以便不会下降。如果杆子倾斜超过 15°或者推车从中心移动超过 2.4 个单位，则游戏结束。OpenAI.com 的主页是这样介绍这个游戏的：

这种环境的小尺寸和简单性保证了能够进行非常快速的实验，这是学习基础知识时必不可少的（*The small size and simplicity of this environment make it possible to run very quick experiments, which is essential when learning the basics*）。

游戏只有四个观察和两个动作，动作是通过施加 +1 或 -1 的力量来移动小车。观察结果包括推车的位置、推车的速度、杆子的角度以及杆子的旋转速度。然而，学习观察语义的知识不是学习最大化游戏奖励所必需的。

现在加载一个流行的游戏环境 CartPole-v0，然后用随机方式播放它：

1）使用标准 make 函数创建 env 对象：

```
env = gym.make('CartPole-v0')
```

2）片段 (episodes) 的数量是游戏的数量，现在将它设置为 1，表示只想玩一次游戏。由于每个片段都是随机的，因此在实际的制作过程中，将运行多个片段并计算奖励的平均值。此外，还可以初始化一个数组来保存在每个时间步长的环境可视化：

```
n_episodes = 1
env_vis = []
```

3）运行两个嵌套循环——外部循环针对片段数量，内部循环针对想要模拟的时间步数。可以一直运行内部循环，直到方案完成或将步数设置为更高的值。

- 在每个片段的开头，使用 env.reset( ) 重置环境。
- 在每个时间步长的开头，使用 env.render( ) 实现可视化。

```
for i_episode in range(n_episodes):
    observation = env.reset()
    for t in range(100):
        env_vis.append(env.render(mode = 'rgb_array'))
        print(observation)
        action = env.action_space.sample()
        observation, reward, done, info = env.step(action)
        if done:
            print("Episode finished at t{}".format(t+1))
            break
```

4）用辅助函数渲染环境：

```
env_render(env_vis)
```

5）辅助函数的代码如下：

```
def env_render(env_vis):
    plt.figure()
    plot = plt.imshow(env_vis[0])
    plt.axis('off')
    def animate(i):
        plot.set_data(env_vis[i])

    anim = anm.FuncAnimation(plt.gcf(),
                             animate,
                             frames=len(env_vis),
                             interval=20,
                             repeat=True,
                             repeat_delay=20)
    display(display_animation(anim, default_mode='loop'))
```

运行此示例时，会得到以下的输出结果：

```
[-0.00666995 -0.03699492 -0.00972623  0.00287713]
[-0.00740985  0.15826516 -0.00966868 -0.29285861]
[-0.00424454 -0.03671761 -0.01552586 -0.00324067]
[-0.0049789  -0.2316135  -0.01559067  0.28450351]
[-0.00961117 -0.42650966 -0.0099006   0.57222875]
[-0.01814136 -0.23125029  0.00154398  0.27644332]
[-0.02276636 -0.0361504   0.00707284 -0.01575223]
[-0.02348937  0.1588694   0.0067578  -0.30619523]
[-0.02031198 -0.03634819  0.00063389 -0.01138875]
[-0.02103895  0.15876466  0.00040612 -0.3038716 ]
[-0.01786366  0.35388083 -0.00567131 -0.59642642]
[-0.01078604  0.54908168 -0.01759984 -0.89089036]
[  1.95594914e-04   7.44437934e-01  -3.54176495e-02  -1.18905344e+00]
[ 0.01508435  0.54979251 -0.05919872 -0.90767902]
[ 0.0260802   0.35551978 -0.0773523  -0.63417465]
[ 0.0331906   0.55163065 -0.09003579 -0.95018025]
[ 0.04422321  0.74784161 -0.1090394  -1.26973934]
[ 0.05918004  0.55426764 -0.13443418 -1.01309691]
[ 0.0702654   0.36117014 -0.15469612 -0.76546874]
[ 0.0774888   0.16847818 -0.1700055  -0.52518186]
[ 0.08085836  0.3655333  -0.18050913 -0.86624457]
[ 0.08816903  0.56259197 -0.19783403 -1.20981195]
Episode finished at t22
```

杆子需要 22 个时间步长才能变得不平衡,每次运行时,都会得到一个不同的时间步长值,这是由于使用 env.action_space.sample( ) 进行随机选择的结果。

由于该游戏会很迅速地带来损失,随机选择一个动作并应用可能不是最好的策略。对于更长的时间步长,有许多算法可以使杆子保持笔直,例如爬山(Hill Climbing)、随机搜索(Random Search)以及策略梯度(Policy Gradient)。

一些解决 Cartpole 游戏的算法可通过以下链接获得:
https://openai.com/requests-for-research/#cartpole.
http://kvfrans.com/simple-algoritms-for-solving-cartpole/.
https://github.com/kvfrans/openai-cartpole。

13.2 将简单的策略应用于 cartpole 游戏

到目前为止,已经随机选择并应用了一个动作。现在采用一些逻辑来挑选动作,而不是随机地进行。第三个观察指的是角度,如果角度大于零,则意味着杆子向右倾斜,因此需要将推车向右移动(1)。否则,将推车向左移动(0)。下面介绍一个例子:

1)定义如下两个策略功能:

```
def policy_logic(env,obs):
    return 1 if obs[2] > 0 else 0
def policy_random(env,obs):
    return env.action_space.sample()
```

2)接下来,定义一个实验函数,该函数将针对特定数量的片段运行,每个片段都会一直运行，直到游戏失败（即 done 为 True）。使用 rewards_max 指示何时停止循环，因为不想让实验一直运行下去：

```python
def experiment(policy, n_episodes, rewards_max):
    rewards=np.empty(shape=(n_episodes))
    env = gym.make('CartPole-v0')
    for i in range(n_episodes):
        obs = env.reset()
        done = False
        episode_reward = 0
        while not done:
            action = policy(env,obs)
            obs, reward, done, info = env.step(action)
            episode_reward += reward
            if episode_reward > rewards_max:
                break
        rewards[i]=episode_reward
    print('Policy:{}, Min reward:{}, Max reward:{}'
        .format(policy.__name__,
                min(rewards),
                max(rewards)))
```

3）运行 100 次实验，或直到奖励小于或等于 rewards_max（它的值设置为 10 000）：

```python
n_episodes = 100
rewards_max = 10000
experiment(policy_random, n_episodes, rewards_max)
experiment(policy_logic, n_episodes, rewards_max)
```

可以看到逻辑选择的动作比随机选择的动作更好，但效果没有提升太多：

```
Policy:policy_random, Min reward:9.0, Max reward:63.0, Average reward:20.26
Policy:policy_logic, Min reward:24.0, Max reward:66.0, Average reward:42.81
```

现在进一步通过基于参数修改选择动作的过程。参数将乘以观察值，并且将基于乘法结果（是 0 还是 1）选择动作。采用随机初始化参数的方式修改随机搜索方法。代码如下：

```python
def policy_logic(theta,obs):
    # 只忽略 theta
    return 1 if obs[2] > 0 else 0

def policy_random(theta,obs):
    return 0 if np.matmul(theta,obs) < 0 else 1

def episode(env, policy, rewards_max):
    obs = env.reset()
    done = False
    episode_reward = 0
    if policy.__name__ in ['policy_random']:
        theta = np.random.rand(4) * 2 - 1
    else:
        theta = None
    while not done:
```

```
            action = policy(theta,obs)
            obs, reward, done, info = env.step(action)
            episode_reward += reward
            if episode_reward > rewards_max:
                break
        return episode_reward
    def experiment(policy, n_episodes, rewards_max):
        rewards=np.empty(shape=(n_episodes))
        env = gym.make('CartPole-v0')
        for i in range(n_episodes):
            rewards[i]=episode(env,policy,rewards_max)
            #print("Episode finished at t{}".format(reward))
        print('Policy:{}, Min reward:{}, Max reward:{}, Average reward:{}'
              .format(policy.__name__,
                      np.min(rewards),
                      np.max(rewards),
                      np.mean(rewards)))

n_episodes = 100
rewards_max = 10000
experiment(policy_random, n_episodes, rewards_max)
experiment(policy_logic, n_episodes, rewards_max)
```

可以看到随机搜索确实改善了结果：

```
Policy:policy_random, Min reward:8.0, Max reward:200.0, Average
reward:40.04
Policy:policy_logic, Min reward:25.0, Max reward:62.0, Average
reward:43.03
```

通过随机搜索，改进了结果并获得 200 的最大奖励。但通随机搜索的奖励通常比较低，这是因为随机搜索会尝试各种不良参数，导致整体结果降低。也可以从所有运行中选择最佳参数，然后在生产中使用最佳参数。将代码修改为首先训练参数：

```
    def policy_logic(theta,obs):
        # 只忽略 theta
        return 1 if obs[2] > 0 else 0

    def policy_random(theta,obs):
        return 0 if np.matmul(theta,obs) < 0 else 1

    def episode(env,policy, rewards_max,theta):
        obs = env.reset()
        done = False
        episode_reward = 0

        while not done:
            action = policy(theta,obs)
            obs, reward, done, info = env.step(action)
            episode_reward += reward
            if episode_reward > rewards_max:
                break
        return episode_reward
```

```python
def train(policy, n_episodes, rewards_max):

    env = gym.make('CartPole-v0')
    theta_best = np.empty(shape=[4])
    reward_best = 0

    for i in range(n_episodes):
        if policy.__name__ in ['policy_random']:
            theta = np.random.rand(4) * 2 - 1
        else:
            theta = None
        reward_episode=episode(env,policy,rewards_max, theta)
        if reward_episode > reward_best:
            reward_best = reward_episode
            theta_best = theta.copy()
    return reward_best,theta_best
def experiment(policy, n_episodes, rewards_max, theta=None):
    rewards=np.empty(shape=[n_episodes])
    env = gym.make('CartPole-v0')
    for i in range(n_episodes):
        rewards[i]=episode(env,policy,rewards_max,theta)
        #print("Episode finished at t{}".format(reward))
    print('Policy:{}, Min reward:{}, Max reward:{}, Average reward:{}'
        .format(policy.__name__,
            np.min(rewards),
            np.max(rewards),
            np.mean(rewards)))

n_episodes = 100
rewards_max = 10000

reward,theta = train(policy_random, n_episodes, rewards_max)
print('trained theta: {}, rewards: {}'.format(theta,reward))
experiment(policy_random, n_episodes, rewards_max, theta)
experiment(policy_logic, n_episodes, rewards_max)
```

训练了 100 个片段，然后使用最佳参数为随机搜索策略运行实验：

```
n_episodes = 100
rewards_max = 10000

reward,theta = train(policy_random, n_episodes, rewards_max)
print('trained theta: {}, rewards: {}'.format(theta,reward))
experiment(policy_random, n_episodes, rewards_max, theta)
experiment(policy_logic, n_episodes, rewards_max)
```

可看到训练参数给出了 200 的最佳结果：

```
trained theta: [-0.14779543  0.93269603  0.70896423  0.84632461],
rewards:200.0
Policy:policy_random, Min reward:200.0, Max reward:200.0, Average
reward:200.0
Policy:policy_logic, Min reward:24.0, Max reward:63.0, Average
reward:41.94
```

可以优化训练代码继续训练,直到获得最大奖励。

在 Jupyter 笔记本 ch-13a_Reinforcement_Learning_NN 中可找到优化过的代码。

现在已经学习了 OpenAI Gym 的基础知识,下面将学习强化学习。

13.3　强化学习 101

可以这样描述强化学习的原理:由代理从前一时间步骤中获得观察和奖励的输入,并将产生的输出作为以最大化累积奖励为目标的动作。

代理包含策略、价值函数和模型:

- 代理用于选择下一个操作的算法称为**策略**。在上一节中,编写了一个策略,它会取一组参数 theta,并根据观察和参数相乘来返回下一个动作。该策略由下式表示:

$$\pi(s): S \rightarrow A,$$

其中,S 是状态集;A 是一组动作。

- 策略是确定性的或随机的。
    - 确定性策略在每次运行中为同一状态返回相同的操作:

    $$\pi(s)=a$$

    - 随机策略在每次运行时对相同的状态、相同的操作返回不同的概率:

    $$\pi(a|s)=P(A=a|S=s)$$

- **价值函数**(value function)会根据当前状态中所选动作预测长期奖励量。因此,价值函数与代理使用的策略有关。奖励表示行动的直接收益,而价价值函数表示动作的累积或将来长期的收益。奖励由环境返回,价值函数由代理在每个时间步骤上估计。

- **模型**表示代理在内部保留的环境,模型可能是环境的不完美表示,代理使用模型估计所选动作的奖励和下一个状态。

代理的目标还可以是为马尔可夫决策过程(Markovian Decision Process,MDP)找到最优策略。MDP 是从一个状态到另一个状态的观察、行动、奖励和过渡的数学表示。为简洁起见,本书将不讨论 MDP,建议好奇的读者在互联网上搜索资源,以深入了解 MDP。

13.3.1　Q 函数(在模型无效时学习优化)

如果模型不可用,则代理通过反复试验学习模型和最优策略,这时代理会使用 Q 函数,该函数的定义如下:

$$Q: S \times A \rightarrow \mathbb{R}$$

如果状态 s 处的代理选择动作 a,则 Q 函数基本上将状态和动作对(pair)映射成表示预期总奖励的实数。

13.3.2　强化学习算法的探索与开发

在没有模型的情况下,代理在每一步都要探索或利用。**探索**意味着代理选择未知动作以找出奖励和模型，**利用**（exploitation）意味着代理会选择最著名的动作以获得最大奖励。如果代理总是决定利用，那么可能会陷入局部最优值。因此，有时代理会绕过学到的策略探索未知的行为。同样，如果代理总是决定探索，那么可能无法找到最优策略。因此，在探索和利用之间取得平衡非常重要。在代码中，为了实现这一目标，可通过使用概率 p 选择随机动作，用概率 $1-p$ 选择最优动作。

13.3.3 V 函数（在模型可用时学习优化）

如果事先已知模型，则代理可以执行策略搜索以找到最大化价值函数的最优策略。当模型可用时，代理使用价值函数，该函数可以定义为未来状态的奖励总和：

$$V^{\pi}(s) = \sum_i R_i \quad \forall s \in S$$

因此，在时间步长 t 处，对使用策略 p 选择操作的值将是：

$$V_t^{\pi} = R_t + R_{t+1} + \cdots + R_{t+n}$$

其中，V 是值；R 是奖励；价值函数估计最多只能达到 n 个时间步长。

当代理使用这种方法估算奖励时，会平等地将所有行为视为奖励。在小推车的示例中，如果在步骤 50 步时进行投票（poll），则代理将把这 50 步视为对失败(fall)负有相同的责任。因此，不是添加未来奖励，而是估计将来奖励的加权总和。通常，权重是提高时间步长强度的折扣率（discount rate），如果折扣率为零，则价值函数变为上面讨论的朴素函数，并且如果折扣率接近 1（例如 0.9 或 0.92），那么与当前奖励相比，未来奖励的影响较小。

因此，现在动作 a 在时间步长 t 处的值将是：

$$V_t^{\pi} = R_t + r \times R_{t+1} = R_t + rR_{t+1} + r^2 R_{t+2} + \cdots + r^n R_{t+n}$$

其中，V 是值；R 是奖励；r 是折扣率。

V 函数和 Q 函数之间的关系：
$V^*(s)$ 是状态 s 的最优价值函数，给出最大奖励；$Q^*(s, a)$ 是状态 s 的最佳 Q 函数，通过选择动作 a 给出最大期望奖励。因此，$V^*(s)$ 是所有可能动作的最优 Q 函数 $Q^*(s, a)$ 的最大值：

$$V^*(s) = \max_a Q^*(s, a) \quad \forall s \in S$$

13.3.4 强化学习技巧

可以根据模型的可用性对强化学习技术进行分类，如下所示：

• **模型可用**：如果模型可用，则代理可以通过迭代策略或价值函数计划离线，以找到提供最大奖励的最优策略。

  • **价值迭代学习**：在价值迭代学习方法中，代理首先将 $V(s)$ 初始化为随机值，然后反复更新 $V(s)$，直到找到最大奖励。

- **策略迭代学习**：在策略迭代学习方法中，代理通过初始化随机策略 p 开始，然后重复更新策略，直到找到最大奖励。
- **模型不可用**：如果模型不可用，则代理只能通过观察其操作的结果来学习。因此，从观察、行动和奖励的历史来看，代理会尝试估计模型或尝试直接推导出最优策略。
  - **基于模型的学习**：在基于模型的学习中，代理首先从历史中估计模型，然后使用策略或以价值为基础的方法寻找最优策略。
  - **无模型学习**：在无模型学习中，代理不会估计模型，而是直接从历史中估计最优策略。Q-Learning 是无模型学习的一个例子。

下面是价值迭代学习算法的一个例子：

```
initialize V(s) to random values for all states
Repeat
    for s in states
        for a in actions
            compute Q[s,a]
        V(s) = max(Q[s])    # 该状态所有操作的最大Q值
Until optimal value of V(s) is found for all states
```

策略迭代学习的算法如下：

```
initialize a policy P_new to random sequence of actions for all states
Repeat
    P = P_new
    for s in states
        compute V(s) with P[s]
        P_new[s] = policy of optimal V(s)
Until P == P_new
```

13.4 强化学习的朴素神经网络策略

按照以下策略进行处理：

1）实现一个朴素的神经网络策略。为了返回动作，定义一个新策略来使用基于预测的神经网络：

```
def policy_naive_nn(nn,obs):
    return np.argmax(nn.predict(np.array([obs])))
```

2）将神经网络定义为一个简单的单层 MLP 网络，将具有 4 个维度的观测值作为输入，并产生 2 个动作的概率：

```
from keras.models import Sequential
from keras.layers import Dense
model = Sequential()
model.add(Dense(8,input_dim=4, activation='relu'))
model.add(Dense(2, activation='softmax'))
model.compile(loss='categorical_crossentropy',optimizer='adam')
model.summary()
```

下面就是模型：

```
Layer (type)                    Output Shape                  Param #
=================================================================
dense_16 (Dense)                (None, 8)                     40
_____
dense_17 (Dense)                (None, 2)                     18
=================================================================
Total params: 58
Trainable params: 58
Non-trainable params: 0
```

3）需要训练这个模型，运行 100 个片段，并只收集分数大于 100 的那些片段的训练数据。如果分数小于 100，那么这些状态和动作不值得记录，因为这不是良好游戏的例子：

```
# 创建训练数据
env = gym.make('CartPole-v0')
n_obs = 4
n_actions = 2
theta = np.random.rand(4) * 2 - 1
n_episodes = 100
r_max = 0
t_max = 0

x_train, y_train = experiment(env,
                              policy_random,
                              n_episodes,
                              theta,r_max,t_max,
                              return_hist_reward=100 )
y_train = np.eye(n_actions)[y_train]
print(x_train.shape,y_train.shape)
```

可以收集 5 732 个样本进行训练：

```
(5732, 4) (5732, 2)
```

4）接下来，训练这个模型：

```
model.fit(x_train, y_train, epochs=50, batch_size=10)
```

5）训练得到的模型可用于玩游戏。但是，需要加入更新训练数据的循环，模型才会进一步从游戏中学习：

```
n_episodes = 200
r_max = 0
t_max = 0

_ = experiment(env,
               policy_naive_nn,
               n_episodes,
               theta=model,
               r_max=r_max,
               t_max=t_max,
               return_hist_reward=0 )

_ = experiment(env,
```

```
            policy_random,
            n_episodes,
            theta,r_max,t_max,
            return_hist_reward=0 )
```

从上面的代码可以看出这种简单的策略几乎以同样的方式执行，但结果比随机策略好一点：

```
Policy:policy_naive_nn, Min reward:37.0, Max reward:200.0, Average
reward:71.05
Policy:policy_random, Min reward:36.0, Max reward:200.0, Average
reward:68.755
```

> 可以通过调整网络和超参数，或通过学习更多游戏玩法进一步改善结果。但是还有更好的算法(例如 Q-Learning) 可以使用。

在本章的剩余部分，将重点关注 Q-Learning 算法，因为大多数现实生活中的问题涉及无模型（mode-free）学习。

13.5 实施 Q-Learning

Q-Learning 是一种无模型方法，可以找到最大化代理奖励的最优策略。在最初的游戏过程中，代理会为每个（状态，动作）对学习 Q 值，也称为探索策略，如前面部分所述。一旦学习了 Q 值，那么最优策略将是在每个状态中选择具有最大 Q 值的动作，这称为利用策略。学习算法可以在找到局部最优解后就结束，因此需要设置 exploration_rate 参数来使用探索策略。

Q-Learning 策略如下：

```
    initialize Q(shape=[#s,#a]) to random values or zeroes
    Repeat (for each episode)
       observe current state s
       Repeat
          select an action a (apply explore or exploit strategy)
          observe state s_next as a result of action a
          update the Q-Table using bellman's equation
          set current state s = s_next
    until the episode ends or a max reward / max steps condition is
reached Until a number of episodes or a condition is reached
          (such as max consecutive wins)
```

前面算法中的 $Q(s,a)$ 表示在前面部分中描述的 Q 函数，该函数的值用于选择操作而不是奖励，因此该函数代表奖励或折扣奖励。此外，使用未来状态中 Q 函数的值更新 Q 函数的值。通过著名的 bellman 方程得到更新：

$$Q(s_t, a_t) = r_t + \gamma \max_a Q(s_{t+1}, a)$$

这基本上意味着在时间步长 t、状态 s、动作 a 下，最大的将来奖励（Q）等于来自当前状态的奖励加上来自下一个状态的最大将来奖励。

$Q(s,a)$ 可以实现为 Q 表（Q-Table）或称为神经网络（Q 网络）。在这两种情况下，

Q 表或 Q 网络的任务是基于给定输入的 Q 值提供最有可能的动作。随着 Q 表变大，基于 Q 表的方法通常变得很困难，因此神经网络成为 Q 网络逼近 Q 函数的最佳选择者。下面来看看这两种方法的实际应用。

 可以在 Jupyter 笔记本 ch-13b_Reinforcement_Learning_DQN 中找到相关代码。

13.5.1 Q-Learning 的初始化和离散化

小车环境返回的观察涉及环境状况。小车状态需要使用离散的连续值表示。

如果将这些值离散化为小的状态空间，那么代理会得到更快的训练，但需要注意的是这会有收敛到最优策略的风险。

使用以下辅助函数离散小车环境的状态空间：

```
# 将值离散化为状态空间
def discretize(val,bounds,n_states):
    discrete_val = 0
    if val <= bounds[0]:
        discrete_val = 0
    elif val >= bounds[1]:
        discrete_val = n_states-1
    else:
        discrete_val = int(round( (n_states-1) *
                                  ((val-bounds[0])/
                                  (bounds[1]-bounds[0]))
                                ))
    return discrete_val

def discretize_state(vals,s_bounds,n_s):
    discrete_vals = []
    for i in range(len(n_s)):
        discrete_vals.append(discretize(vals[i],s_bounds[i],n_s[i]))
    return np.array(discrete_vals,dtype=np.int)
```

将每个观察维度空间离散为 10 个单位，也可以尝试不同的离散化空间。在离散化之后，找到观察的上限和下限，并将速度和角速度的界限改为介于 −1 和 +1 之间，而不是 −Inf 和 +Inf 之间。代码如下：

```
env = gym.make('CartPole-v0')
n_a = env.action_space.n
# 每个观察维度的离散状态数量
n_s = np.array([10,10,10,10])    # 位置、速度、角度、角速度
s_bounds = np.array(list(zip(env.observation_space.low,
env.observation_space.high)))
# 速度和角速度界限太高, 故绑定在-1和+1之间
s_bounds[1] = (-1.0,1.0)
s_bounds[3] = (-1.0,1.0)
```

13.5.2 基于 Q 表的 Q-Learning

可以在 ch-13b.ipynb 中按照本节的代码进行操作。由于离散空间的维度为 [10,10,10,10]，因此 Q 表的大小为 [10,10,10,10,2]：

```
# 创建一个形状为(10,10,10,10, 2)的Q表表示S X A -> R
q_table = np.zeros(shape = np.append(n_s,n_a))
```

定义一个 Q 表策略，它是基于 exploration_rate 参数的探索或利用：

```
def policy_q_table(state, env):
    # 探索策略——选择一个随机动作
    if np.random.random() < explore_rate:
        action = env.action_space.sample()
    # 利用策略——选择q最高的动作
    else:
        action = np.argmax(q_table[tuple(state)])
    return action
```

定义运行单个片段的 episode() 函数，如下所示：

1）从初始化变量和第一个状态开始：

```
obs = env.reset()
state_prev = discretize_state(obs,s_bounds,n_s)

episode_reward = 0
done = False
t = 0
```

2）选择动作并观察下一个状态：

```
action = policy(state_prev, env)
obs, reward, done, info = env.step(action)
state_new = discretize_state(obs,s_bounds,n_s)
```

3）更新 Q 表：

```
best_q = np.amax(q_table[tuple(state_new)])
bellman_q = reward + discount_rate * best_q
indices = tuple(np.append(state_prev,action))
q_table[indices] += learning_rate*( bellman_q - q_table[indices])
```

4）将下一个状态设置为上一个状态，并将奖励添加到片段的奖励中：

```
state_prev = state_new
episode_reward += reward
```

experiment() 函数调用 episode 函数并累积报告的奖励，希望修改该函数以检查连续获胜，以及其他跟具体的游戏有关的逻辑：

```
# 收集每个片段的观察和奖励
def experiment(env, policy, n_episodes,r_max=0, t_max=0):
    rewards=np.empty(shape=[n_episodes])
    for i in range(n_episodes):
        val = episode(env, policy, r_max, t_max)
        rewards[i]=val
    print('Policy:{}, Min reward:{}, Max reward:{}, Average reward:{}'
```

```
          .format(policy.__name__,
                  np.min(rewards),
                  np.max(rewards),
                  np.mean(rewards)))
```

现在,要做的就是定义参数(例如 learning_rate、discount_rate 和 explore_rate),并运行 experiment( ) 函数,如下所示:

```
learning_rate = 0.8
discount_rate = 0.9
explore_rate = 0.2
n_episodes = 1000
experiment(env, policy_q_table, n_episodes)
```

采用简单实现方法,在 1000 个片段的情况下,基于 Q 表策略的最大奖励为 180:

```
Policy:policy_q_table, Min reward:8.0, Max reward:180.0, Average
reward:17.592
```

对算法的实现很容易解释。但是,还可以修改代码将探索率的初始值设置得高一些,然后随着时间步长而衰减。同样,也可以实现学习和折扣率的衰减逻辑。下面来看看当 Q 函数学得更快时,是否可以用更少的片段获得更高的奖励。

13.5.3 使用 Q 网络或深度 Q 网络(DQN)进行 Q-Learning

在 DQN 中,将 Q 表替换为神经网络(Q 网络),当使用探索的状态及其 Q 值连续训练时,将学习具有最佳动作的响应。因此,为了训练网络,需要一个存储游戏的内存:

1)使用大小为 1000 的双端队列实现游戏内存:

```
memory = deque(maxlen=1000)
```

2)接下来,构建一个简单且具有隐藏层的神经网络模型 q_nn:

```
from keras.models import Sequential
from keras.layers import Dense
model = Sequential()
model.add(Dense(8,input_dim=4, activation='relu'))
model.add(Dense(2, activation='linear'))
model.compile(loss='mse',optimizer='adam')
model.summary()
q_nn = model
```

Q 网络的结构如下所示:

```
Layer (type)                 Output Shape              Param #
=================================================================
dense_1 (Dense)              (None, 8)                 40
_____
dense_2 (Dense)              (None, 2)                 18
=================================================================
Total params: 58
Trainable params: 58
Non-trainable params: 0
_____
```

执行游戏片段的 episode( ) 函数包含了对基于 Q 网络的算法进行的如下改进：

1）生成下一个状态后，将状态、动作和奖励添加到游戏内存中：

```
action = policy(state_prev, env)
obs, reward, done, info = env.step(action)
state_next = discretize_state(obs,s_bounds,n_s)
# 将 state_prev、action、reward、state_new、done 添加到内存中
memory.append([state_prev,action,reward,state_next,done])
```

2）使用 bellman 函数生成并更新具有最大将来奖励的 q_values：

```
states = np.array([x[0] for x in memory])
states_next = np.array([np.zeros(4) if x[4] else x[3] for x in memory])
q_values = q_nn.predict(states)
q_values_next = q_nn.predict(states_next)

for i in range(len(memory)):
    state_prev,action,reward,state_next,done = memory[i]
    if done:
        q_values[i,action] = reward
    else:
        best_q = np.amax(q_values_next[i])
        bellman_q = reward + discount_rate * best_q
        q_values[i,action] = bellman_q
```

3）使用从内存中收到的状态和 q_values 训练 q_nn：

```
q_nn.fit(states,q_values,epochs=1,batch_size=50,verbose=0)
```

深度强化学习文献将游戏玩法保存在内存中，并使用它来训练模型的过程称为**记忆重放**（memory replay）。按照以下方式运行基于 DQN 的游戏：

```
learning_rate = 0.8
discount_rate = 0.9
explore_rate = 0.2
n_episodes = 100
experiment(env, policy_q_nn, n_episodes)
```

这会获得 150 的最大奖励。可以通过超参数调整、网络调整以及使用折扣率和探索率的速率衰减来改进：

```
Policy:policy_q_nn, Min reward:8.0, Max reward:150.0, Average reward:41.27
```

现在可以在每一步计算和训练模型；读者可以去研究如何在片断之后将其更改为训练。此外，可以更改代码以丢弃内存重放，并为返回较小奖励的片段重新训练模型。但是，请谨慎实现此选项，因为可能会减慢学习速度，这是因为初始游戏会更频繁地产生较小的奖励。

13.6 总结

本章学习了如何在 Keras 中实现强化学习算法。为了让示例变得简单，这里采用了

Keras，也可以使用 TensorFlow 实现相同的网络和模型。本章只使用了单层 MLP，因为示例游戏非常简单，但对于复杂的示例，最终可能会使用复杂的 CNN、RNN 或序列到序列的模型。

本章还介绍了 OpenAI Gym，这是一个框架，该框架提供了一个模拟许多流行游戏的环境，以实现和实践强化学习算法。本章介绍了深度强化学习的概念，鼓励读者阅读强化学习的专业书籍，以便能更深入地研究相关的理论和概念。

强化学习是一种先进的技术，常用于解决复杂的问题。下一章将学习另一种先进的深度学习技术——生成对抗网络。

第 14 章
生成对抗网络（GAN）

训练生成模型是为了生成与训练数据类似的更多数据，而训练对抗模型是为了通过提供对抗性示例来区分真实数据和假数据。

生成对抗网络（GAN） 结合了两种模型的特征。GAN 由两部分组成：
- 生成模型：可以学习如何生成类似的数据；
- 判别模型：用于学习如何区分真实数据和生成数据（来自生成模型）。

GAN 已成功应用于各种复杂问题，例如：
- 从低分辨率图像中生成能逼近真实分辨率的图像；
- 从文本中合成图像；
- 样式迁移；
- 补充不完整的图像和视频。

本章为了介绍如何在 TensorFlow 和 Keras 中实现 GAN，将讨论以下主题：
- GAN 101；
- 基于 TensorFlow 的简单 GAN；
- 基于 Keras 的简单 GAN；
- 基于 TensorFlow 和 Keras 的深度卷积 GAN。

14.1 GAN 101

如图 14-1 所示，生成对抗网络（GAN）有两个同步工作的模型，它们是在复杂数据（如图像、视频或音频文件）上进行学习和训练：

生成模型从随机噪声开始生成数据，然后慢慢地学习如何生成更真实的数据。生成器的输出结果和真实数据被输入给判别器，让判别器学习如何区分假数据和真实数据。

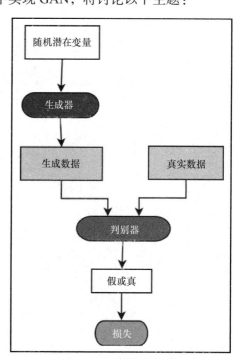

图 14-1

第14章
生成对抗网络（GAN）

因此，生成器和判别器相互进行对抗性游戏 (game)，其中生成器试图尽可能生成与真实数据一样的数据来欺骗判别器，判别器试图通过从真实数据中识别伪造的数据来避免被欺骗，因此判别器试图最小化分类损失函数。两种模型都以锁步（lockstep）方式进行训练。

在数学上，生成模型 $G(z)$ 会学习概率分布 $p(z)$，使判别器 $D(G(z), x)$ 不能识别在 $p(z)$ 和 $p(x)$ 之间的概率分布，GAN 的目标函数可以通过以下价值函数 V 来描述（来自 https://papers.nips.cc/paper/5423-generative-adversarial-nets.pdf）：

$$\min_G \max_D V(D,G) = E_{x \sim p_{data}(x)}[\log D(x)] + E_{z \sim p_z(z)}[\log(1 - D(G(z)))]$$

可以在以下链接中找到 Ian Goodfellow 在 NIPS 2016 上关于 GAN 的开创性教程：https://arxiv.org/pdf/1701.00160.pdf。

这里描述了一个简单的 GAN（在文献中也称为 vanilla GAN），由 Ian Goodfellow 在他的开创性论文 (https://arxiv.org/abs/1406.2661) 中首次介绍这种 GAN 模型。从那时起，学者们对 GAN 进行了广泛的研究，推导出不同架构，并应用于不同的领域。

例如，在条件 GAN 中，为生成器和判别器网络提供标签，使得条件 GAN 的目标函数可以通过如下的价值函数 V 进行等价描述：

$$\min_G \max_D V(D,G) = E_{x \sim p_{data}(x)}[\log D(x)] + E_{z \sim p_z(z)}[\log(1 - D(G(z,y),y))]$$

介绍条件 GAN 的原始论文的下载链接为：https://arxiv.org/abs/1411.1784。

在一些应用程序中，对 GAN 进行了改进，如文本到图像、图像合成、图像标记、样式迁移和图像迁移等，具体的改进版本如表 14-1 所示。

表 14-1

GAN 的改进版本	原始论文	应用领域
StackGAN	https://arxiv.org/abs/1710.10916	文字图像
StackGAN++	https://arxiv.org/abs/1612.03242	逼真的图像合成
DCGAN	https://arxiv.org/abs/1511.06434	图像合成
HR-DCGAN	https://arxiv.org/abs/1711.06491	高分辨率图像合成
条件 GAN	https://arxiv.org/abs/1411.1784	图像标记
InfoGAN	https://arxiv.org/abs/1606.03657	风格识别
Wasserstein GAN	https://arxiv.org/abs/1701.07875 https://arxiv.org/abs/1704.00028	图像生成
Coupled GAN	https://arxiv.org/abs/1606.07536	图像转换，域适应
BE GAN	https://arxiv.org/abs/1703.10717	图像生成
DiscoGAN	https://arxiv.org/abs/1703.05192	样式迁移
CycleGAN	https://arxiv.org/abs/1703.10593	样式迁移

下面将使用 MNIST 数据集创建一个简单的 GAN。在本练习中，会使用以下函数将

MNIST 数据集的值归一化在 [-1, +1] 之间：

```
def norm(x):
    return (x-0.5)/0.5
```

另外定义 256 维的随机噪声，用于测试生成模型：

```
n_z = 256
z_test = np.random.uniform(-1.0,1.0,size=[8,n_z])
```

下面这个函数将显示生成的图像，该函数会在本章所有示例中使用：

```
def display_images(images):
    for i in range(images.shape[0]):
        plt.subplot(1, 8, i + 1)
        plt.imshow(images[i])
        plt.axis('off')
    plt.tight_layout()
    plt.show()
```

14.2 建立和训练 GAN 的最佳实践

对于用于演示的数据集，判别器能对真实图像和假图像进行非常好的分类，因此不会向生成器提供大量的梯度反馈。因此，必须通过下面的方式使判别器变弱：

- 让判别器的学习率远高于生成器的学习率。
- 判别器的优化器是 GradientDescent，生成器的优化器是 Adam。
- 判别器采用 dropout 正则化，而生成器则没有进行正则化。
- 与生成器相比，判别器具有更少的层和更少的神经元。
- 生成器的输出对应 tanh 函数，而判别器的输出对应 sigmoid 函数。
- 在 Keras 模型中，对于真实数据的标签，使用 0.9 而不是 1.0，而对于伪造数据的标签使用 0.1 而不是 0.0，这样做是为了在标签中引入一点噪声。欢迎读者研究并尝试其他的最佳实践方案。

14.3 基于 TensorFlow 的简单 GAN

 可以查看 Jupyter 笔记本 ch-14a_SimpleGAN 中的代码。

为了使用 TensorFlow 构建 GAN，使用以下步骤构建三个网络、两个判别器模型和一个生成器模型：

1）首先添加用于定义网络的超参数：

```
# 图形超参数
g_learning_rate = 0.00001
d_learning_rate = 0.01
n_x = 784  # MNIST 图像中的像素数
```

```python
# 生成器和判别器的隐藏层数
g_n_layers = 3
d_n_layers = 1
# 每个隐藏层中的神经元
g_n_neurons = [256, 512, 1024]
d_n_neurons = [256]

# 定义参数 ditionary
d_params = {}
g_params = {}

activation = tf.nn.leaky_relu
w_initializer = tf.glorot_uniform_initializer
b_initializer = tf.zeros_initializer
```

2）定义生成器网络

```python
z_p = tf.placeholder(dtype=tf.float32, name='z_p',
        shape=[None, n_z])
layer = z_p

# 添加生成器网络权重、偏差和层
with tf.variable_scope('g'):
    for i in range(0, g_n_layers):
        w_name = 'w_{0:04d}'.format(i)
        g_params[w_name] = tf.get_variable(
            name=w_name,
            shape=[n_z if i == 0 else g_n_neurons[i - 1],
                   g_n_neurons[i]],
            initializer=w_initializer())
        b_name = 'b_{0:04d}'.format(i)
        g_params[b_name] = tf.get_variable(
            name=b_name, shape=[g_n_neurons[i]],
            initializer=b_initializer())
        layer = activation(
            tf.matmul(layer, g_params[w_name]) + g_params[b_name])
    # 输出(logit)层
    i = g_n_layers
    w_name = 'w_{0:04d}'.format(i)
    g_params[w_name] = tf.get_variable(
        name=w_name,
        shape=[g_n_neurons[i - 1], n_x],
        initializer=w_initializer())
    b_name = 'b_{0:04d}'.format(i)
    g_params[b_name] = tf.get_variable(
        name=b_name, shape=[n_x], initializer=b_initializer())
    g_logit = tf.matmul(layer, g_params[w_name]) + g_params[b_name]
    g_model = tf.nn.tanh(g_logit)
```

3）接下来，定义将要构建的两个判别器网络的权重和偏差：

```python
with tf.variable_scope('d'):
    for i in range(0, d_n_layers):
        w_name = 'w_{0:04d}'.format(i)
        d_params[w_name] = tf.get_variable(
```

```
            name=w_name,
            shape=[n_x if i == 0 else d_n_neurons[i - 1],
                d_n_neurons[i]],
            initializer=w_initializer())

        b_name = 'b_{0:04d}'.format(i)
        d_params[b_name] = tf.get_variable(
            name=b_name, shape=[d_n_neurons[i]],
            initializer=b_initializer())

    #输出(logit)层
    i = d_n_layers
    w_name = 'w_{0:04d}'.format(i)
    d_params[w_name] = tf.get_variable(
        name=w_name, shape=[d_n_neurons[i - 1], 1],
        initializer=w_initializer())

    b_name = 'b_{0:04d}'.format(i)
    d_params[b_name] = tf.get_variable(
        name=b_name, shape=[1], initializer=b_initializer())
```

4)现在使用这些参数构建将真实图像作为输入,而该判别器的输出为分类结果:

```
# 定义discriminator_real

# 输入真实图像
x_p = tf.placeholder(dtype=tf.float32, name='x_p',
        shape=[None, n_x])

layer = x_p

with tf.variable_scope('d'):
    for i in range(0, d_n_layers):
        w_name = 'w_{0:04d}'.format(i)
        b_name = 'b_{0:04d}'.format(i)
        layer = activation(
            tf.matmul(layer, d_params[w_name]) + d_params[b_name])
        layer = tf.nn.dropout(layer,0.7)
    #输出(logit)层
    i = d_n_layers
    w_name = 'w_{0:04d}'.format(i)
    b_name = 'b_{0:04d}'.format(i)
    d_logit_real = tf.matmul(layer,
        d_params[w_name]) + d_params[b_name]
    d_model_real = tf.nn.sigmoid(d_logit_real)
```

5)接下来,构建另一个具有相同参数的判别器网络,而将生成器的输出作为输入:

```
# 定义 discriminator_fake

# 输入生成的假图像
z = g_model
layer = z

with tf.variable_scope('d'):
```

```
        for i in range(0, d_n_layers):
            w_name = 'w_{0:04d}'.format(i)
            b_name = 'b_{0:04d}'.format(i)
            layer = activation(
                tf.matmul(layer, d_params[w_name]) + d_params[b_name])
            layer = tf.nn.dropout(layer,0.7)
        # 输出(logit)层
        i = d_n_layers
        w_name = 'w_{0:04d}'.format(i)
        b_name = 'b_{0:04d}'.format(i)
        d_logit_fake = tf.matmul(layer,
            d_params[w_name]) + d_params[b_name]
        d_model_fake = tf.nn.sigmoid(d_logit_fake)
```

6）现在已经构建了三个网络，这些网络之间是通过损失函数、优化器和训练函数进行连接的。在训练生成器时，只训练生成器的参数；当训练判别器时，只训练判别器的参数。使用 var_list 指定这些参数，并将 var_list 以参数形式传递给优化器的 minimize( ) 函数。下面是完整的代码，这些代码定义两种网络的损失函数、优化器和训练函数：

```
        g_loss = -tf.reduce_mean(tf.log(d_model_fake))
        d_loss = -tf.reduce_mean(tf.log(d_model_real) + tf.log(1 -
        d_model_fake))
        g_optimizer = tf.train.AdamOptimizer(g_learning_rate)
        d_optimizer = tf.train.GradientDescentOptimizer(d_learning_rate)

        g_train_op = g_optimizer.minimize(g_loss,
                        var_list=list(g_params.values()))
        d_train_op = d_optimizer.minimize(d_loss,
                        var_list=list(d_params.values()))
```

7）现在已经定义了模型，接下来会训练这些模型。按以下算法实现训练：

```
For each epoch:
    For each batch:
        get real images x_batch
        generate noise z_batch
        train discriminator using z_batch and x_batch
        generate noise z_batch
        train generator using z_batch
```

完整的训练代码如下：

```
        n_epochs = 400
        batch_size = 100
        n_batches = int(mnist.train.num_examples / batch_size)
        n_epochs_print = 50

        with tf.Session() as tfs:
            tfs.run(tf.global_variables_initializer())
            for epoch in range(n_epochs):
                epoch_d_loss = 0.0
                epoch_g_loss = 0.0
                for batch in range(n_batches):
                    x_batch, _ = mnist.train.next_batch(batch_size)
```

```python
        x_batch = norm(x_batch)
        z_batch = np.random.uniform(-1.0,1.0,size=[batch_size,n_z])
        feed_dict = {x_p: x_batch,z_p: z_batch}
        _,batch_d_loss = tfs.run([d_train_op,d_loss],
                                 feed_dict=feed_dict)
        z_batch = np.random.uniform(-1.0,1.0,size=[batch_size,n_z])
        feed_dict={z_p: z_batch}
        _,batch_g_loss = tfs.run([g_train_op,g_loss],
                                 feed_dict=feed_dict)
        epoch_d_loss += batch_d_loss
        epoch_g_loss += batch_g_loss
    if epoch%n_epochs_print == 0:
        average_d_loss = epoch_d_loss / n_batches
        average_g_loss = epoch_g_loss / n_batches
        print('epoch: {0:04d}   d_loss = {1:0.6f}  g_loss = {2:0.6f}'
              .format(epoch,average_d_loss,average_g_loss))
        # 使用已训练的生成器模型预测图像
        x_pred = tfs.run(g_model,feed_dict={z_p:z_test})
        display_images(x_pred.reshape(-1,pixel_size,pixel_size))
```

每 50 个迭代周期输出一次生成的图像，如图 14-2 所示。

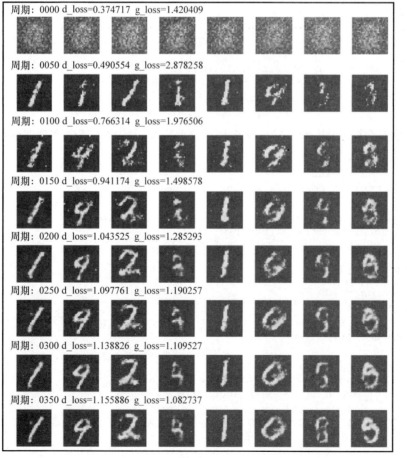

图 14-2

如图14-2所示，生成器在迭代周期为0时产生噪声，但是在迭代周期为350时，经过训练之后可以产生更好的手写数字。可以尝试使用不同的迭代周期、正则化、网络架构和其他超参数进行试验，看看是否可以产生更快更好的结果。

14.4 基于 Keras 的简单 GAN

 可以按照 Jupyter 笔记本 ch-14a_SimpleGAN 中的代码进行操作。

下面介绍如何在 Keras 中实现相同的模型：

1）定义与上一节一样的超参数：

```
# 图形超参数
g_learning_rate = 0.00001
d_learning_rate = 0.01
n_x = 784  # MNIST图像中的像素数量
# 生成器和判别器的隐藏层数
g_n_layers = 3
d_n_layers = 1
# 每个隐藏层中的神经元
g_n_neurons = [256, 512, 1024]
d_n_neurons = [256]
```

2）定义生成器网络：

```
# 定义生成器

g_model = Sequential()
g_model.add(Dense(units=g_n_neurons[0],
                  input_shape=(n_z,),
                  name='g_0'))
g_model.add(LeakyReLU())
for i in range(1,g_n_layers):
    g_model.add(Dense(units=g_n_neurons[i],
                      name='g_{}'.format(i)
                      ))
    g_model.add(LeakyReLU())
g_model.add(Dense(units=n_x, activation='tanh',name='g_out'))
print('Generator:')
g_model.summary()
g_model.compile(loss='binary_crossentropy',
                optimizer=keras.optimizers.Adam(lr=g_learning_rate)
                )
```

下面是生成器模型：

```
Generator:
_____
Layer (type)                 Output Shape              Param #
=================================================================
```

g_0 (Dense)	(None, 256)	65792
leaky_re_lu_1 (LeakyReLU)	(None, 256)	0
g_1 (Dense)	(None, 512)	131584
leaky_re_lu_2 (LeakyReLU)	(None, 512)	0
g_2 (Dense)	(None, 1024)	525312
leaky_re_lu_3 (LeakyReLU)	(None, 1024)	0
g_out (Dense)	(None, 784)	803600

```
Total params: 1,526,288
Trainable params: 1,526,288
Non-trainable params: 0
```

3）在 Keras 示例中，没有像 TensorFlow 示例那样定义两个判别器网络。这里只定义了一个判别器网络，然后将生成器和判别器网络放到 GAN 中。GAN 仅用于训练生成器参数，判别器网络用于训练判别器参数：

```
# 定义判别器

d_model = Sequential()
d_model.add(Dense(units=d_n_neurons[0],
                  input_shape=(n_x,),
                  name='d_0'
                 ))
d_model.add(LeakyReLU())
d_model.add(Dropout(0.3))
for i in range(1,d_n_layers):
    d_model.add(Dense(units=d_n_neurons[i],
                      name='d_{}'.format(i)
                     ))
    d_model.add(LeakyReLU())
    d_model.add(Dropout(0.3))
d_model.add(Dense(units=1, activation='sigmoid',name='d_out'))
print('Discriminator:')
d_model.summary()
d_model.compile(loss='binary_crossentropy',
                optimizer=keras.optimizers.SGD(lr=d_learning_rate)
               )
```

下面是判别器模型：

```
Discriminator:
```

Layer (type)	Output Shape	Param #
d_0 (Dense)	(None, 256)	200960

```
leaky_re_lu_4 (LeakyReLU)      (None, 256)           0
_____
dropout_1 (Dropout)            (None, 256)           0
_____
d_out (Dense)                  (None, 1)             257
=================================================================
Total params: 201,217
Trainable params: 201,217
Non-trainable params: 0
_____
```

4) 下面定义 GAN，并将判别器模型的可训练属性设置为 false，因为 GAN 仅用于训练生成器：

```
# 定义GAN
d_model.trainable=False
z_in = Input(shape=(n_z,),name='z_in')
x_in = g_model(z_in)
gan_out = d_model(x_in)

gan_model = Model(inputs=z_in,outputs=gan_out,name='gan')
print('GAN:')
gan_model.summary()
gan_model.compile(loss='binary_crossentropy',
                  optimizer=keras.optimizers.Adam(lr=g_learning_rate)
                  )
```

GAN 模型如下所示：

```
GAN:
_____
Layer (type)                   Output Shape         Param #
=================================================================
z_in (InputLayer)              (None, 256)          0
_____
sequential_1 (Sequential)      (None, 784)          1526288
_____
sequential_2 (Sequential)      (None, 1)            201217
=================================================================
Total params: 1,727,505
Trainable params: 1,526,288
Non-trainable params: 201,217
_____
```

5）现在已经定义了三个模型，接下来必须训练这些模型。训练按照以下算法进行：

```
For each epoch:
  For each batch:
    get real images x_batch
    generate noise z_batch
    generate images g_batch using generator model
    combine g_batch and x_batch into x_in and create labels y_out
```

```
set discriminator model as trainable
train discriminator using x_in and y_out
generate noise z_batch
set x_in = z_batch and labels y_out = 1
set discriminator model as non-trainable
train gan model using x_in and y_out,
    (effectively training generator model)
```

真实图像和假图像的标签分别设置为 0.9 和 0.1。通常建议对标签进行平滑处理，可以对伪数据选择 0.0~0.3 的随机值，为真实数据选择 0.8~1.0 的随机值。

以下是完整的训练代码：

```
n_epochs = 400
batch_size = 100
n_batches = int(mnist.train.num_examples / batch_size)
n_epochs_print = 50

for epoch in range(n_epochs+1):
    epoch_d_loss = 0.0
    epoch_g_loss = 0.0
    for batch in range(n_batches):
        x_batch, _ = mnist.train.next_batch(batch_size)
        x_batch = norm(x_batch)
        z_batch = np.random.uniform(-1.0,1.0,size=[batch_size,n_z])
        g_batch = g_model.predict(z_batch)
        x_in = np.concatenate([x_batch,g_batch])
        y_out = np.ones(batch_size*2)
        y_out[:batch_size]=0.9
        y_out[batch_size:]=0.1
        d_model.trainable=True
        batch_d_loss = d_model.train_on_batch(x_in,y_out)

        z_batch = np.random.uniform(-1.0,1.0,size=[batch_size,n_z])
        x_in=z_batch
        y_out = np.ones(batch_size)
        d_model.trainable=False
        batch_g_loss = gan_model.train_on_batch(x_in,y_out)

        epoch_d_loss += batch_d_loss
        epoch_g_loss += batch_g_loss
    if epoch%n_epochs_print == 0:
        average_d_loss = epoch_d_loss / n_batches
        average_g_loss = epoch_g_loss / n_batches
        print('epoch: {0:04d}   d_loss = {1:0.6f}   g_loss = {2:0.6f}'
            .format(epoch,average_d_loss,average_g_loss))
        # 使用已训练的生成器模型预测图像
        x_pred = g_model.predict(z_test)
        display_images(x_pred.reshape(-1,pixel_size,pixel_size))
```

每 50 个迭代周期输出一次结果，最多进行 350 个迭代周期，如图 14-3 所示。

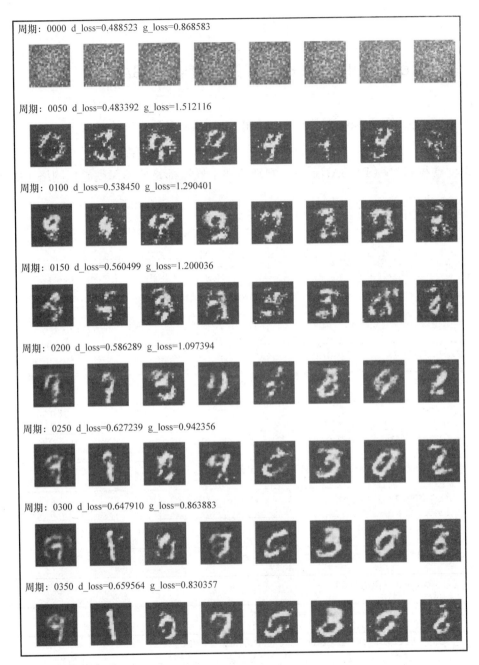

图 14-3

该模型慢慢地从随机噪声中学习生成高质量手写数字图像。

GAN 有很多改进版本,可以写一本书来介绍不同类型的 GAN。但是它们的实现技术基本上与这里介绍的一样。

14.5 基于 TensorFlow 和 Keras 的深度卷积 GAN

 可以按照 Jupyter 笔记本 ch-14b_DCGAN 中的代码进行操作。

在深度卷积 GAN（DCGAN）中，判别器和生成器都是使用深度卷积网络（DCN）实现的。

1）在此示例中生成器会按下面的网络来实现：

```
Generator:

Layer (type)                     Output Shape              Param #
=================================================================
g_in (Dense)                     (None, 3200)              822400
_____
g_in_act (Activation)            (None, 3200)              0
_____
g_in_reshape (Reshape)           (None, 5, 5, 128)         0
_____
g_0_up2d (UpSampling2D)          (None, 10, 10, 128)       0
_____
g_0_conv2d (Conv2D)              (None, 10, 10, 64)        204864
_____
g_0_act (Activation)             (None, 10, 10, 64)        0
_____
g_1_up2d (UpSampling2D)          (None, 20, 20, 64)        0
_____
g_1_conv2d (Conv2D)              (None, 20, 20, 32)        51232
_____
g_1_act (Activation)             (None, 20, 20, 32)        0
_____
g_2_up2d (UpSampling2D)          (None, 40, 40, 32)        0
_____
g_2_conv2d (Conv2D)              (None, 40, 40, 16)        12816
_____
g_2_act (Activation)             (None, 40, 40, 16)        0
_____
g_out_flatten (Flatten)          (None, 25600)             0
_____
g_out (Dense)                    (None, 784)               20071184
=================================================================
Total params: 21,162,496
Trainable params: 21,162,496
Non-trainable params: 0
```

2）判别器是一个更强大的网络，有三个卷积层，激活函数为 tanh 函数。这里将判别器网络定义如下：

```
Discriminator:
Layer (type)                  Output Shape              Param #
=================================================================
d_0_reshape (Reshape)         (None, 28, 28, 1)         0

d_0_conv2d (Conv2D)           (None, 28, 28, 64)        1664

d_0_act (Activation)          (None, 28, 28, 64)        0

d_0_maxpool (MaxPooling2D)    (None, 14, 14, 64)        0

d_out_flatten (Flatten)       (None, 12544)             0

d_out (Dense)                 (None, 1)                 12545
=================================================================
Total params: 14,209
Trainable params: 14,209
Non-trainable params: 0
```

3）GAN 由判别器和生成器组成，具体描述如下：

```
GAN:
Layer (type)                  Output Shape              Param #
=================================================================
z_in (InputLayer)             (None, 256)               0

g (Sequential)                (None, 784)               21162496

d (Sequential)                (None, 1)                 14209
=================================================================
Total params: 21,176,705
Trainable params: 21,162,496
Non-trainable params: 14,209
```

对这个模型运行 400 个迭代周期，会得到如图 14-4 所示的结果。

如图所示，DCGAN 能够在迭代周期为 100 时就开始生成高质量的数字。DCGAN 已被用于样式迁移、图像和标题的生成，以及图像代数（即取一幅图像的一部分并将其添加到另一个图像中）。

在 Jupyter 笔记本 ch-14b_DCGAN 中提供了 MNIST DCGAN 的完整代码。

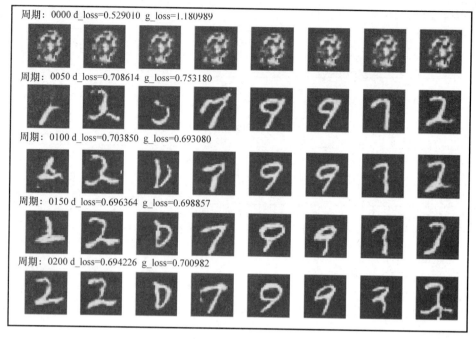

图 14-4

14.6 总结

本章介绍了生成对抗网络（GAN），并介绍如何在 TensorFlow 和 Keras 中构建一个简单的 GAN，以及将其应用在 MNIST 数据集上来生成图像。通过本章的学习，我们了解到许多不同的 GAN 改进版本，例如 DCGAN、SRGAN、StackGAN 和 CycleGAN 等。本章还建立了一个 DCGAN，其中生成器和判别器由深度卷积网络（DCN）组成。鼓励读者尝试不同的改进方法，以找到适合解决具体应用问题的方法。

下一章将介绍如何在分布式集群上使用 TensorFlow 集群和多个计算设备（如多个 GPU）来构建和部署模型。

第 15 章 基于 TensorFlow 集群的分布式模型

前面几章学习了如何在生产中使用 Kubernetes、Docker 和 TensorFlow 运行大规模的 TensorFlow 模型。但 TensorFlow 服务并不是大规模运行 TensorFlow 模型的唯一方法。TensorFlow 还提供了另一种不仅可以运行模型，而且还可以训练模型的机制，这种机制可以在不同节点或不同节点（或同一节点）的不同设备上实现。在第 1 章中还学习了如何放置变量以及如何在不同设备上进行操作。本章将学习如何布署分布式 TensorFlow 模型，这种模型能在多个节点上的多个设备上运行。

本章将介绍以下主题：
- 分布式执行策略；
- TensorFlow 集群；
- 数据并行模型；
- 分布式模型的异步更新和同步更新。

15.1 分布式执行策略

为了在多个设备或节点上布署单个模型的训练，有以下策略：
- **模型并行**：将模型划分为多个计算子图，并将单独的一个计算图放在不同的节点或设备上。子图执行计算并根据需要交换变量。
- **数据并行**：将数据分组，然后在多个节点或设备上运行相同的模型，在主节点上组合参数。因此，工作节点在批量数据上训练模型并将参数更新发送到主节点，也称为参数服务器。

图 15-1 显示了数据并行方法，其中模型复制会分批读取数据分区并将参数更新发送到参数服务器，参数服务器将更新的参数发送回复制模型以便能进行下一个批处理的更新计算。

在 TensorFlow 中，有两种方法能基于数据并行策略来实现在多个节点/设备上进行模型复制：
- **计算图形内复制**：在此方法中，有一个客户端任务拥有模型参数，并将模型的计算任务分配给多个工作器 (worker) 任务。

- **计算图之间复制**：在此方法中，每个客户端任务都连接到自己的工作器以分担模型的计算任务，但所有工作器都更新相同的共享模型。在此模型中，TensorFlow 会自动将一个工作器指定为主要工作器，以便模型参数仅由主要工作器初始化一次。

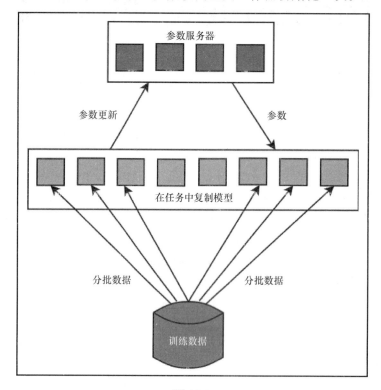

图 15-1

在这两种方法中，参数服务器上的参数可以通过两种不同的方式来更新：

- **同步更新**：在同步更新中，参数服务器进行梯度更新之前等待接收所有工作程序的更新。参数服务器聚合这种更新，例如通过计算所有聚合的平均值并将其应用于参数。更新后，参数将同时发送给所有工作器。这种方法的缺点是一个慢的工作器可能会减慢每个工作器的更新速度。
- **异步更新**：在异步更新中，工作器在准备好后将更新发送到参数服务器，然后参数服务器在接收更新时应用更新并将其发回。这种方法的缺点是，当工作器计算参数并发回更新时，参数可能已被其他工作器多次更新。这个问题可以通过几种方法解决，例如降低批量数据的大小或降低学习率。令人惊讶的是，异步方法居然可以工作，而且这种方法在实际应用中确实有效！

15.2　TensorFlow 集群

TensorFlow（TF）集群是一种实现分布式策略（上一节才讨论过）的机制。在逻辑级别，TF 集群运行一个或多个作业，每个作业由一个或多个任务组成。因此，作业只是任务的逻辑分组。在进程级别，每个任务都作为 TF 服务器运行。在机器级别，每个物理机或节点可以通过运行多个服务器（每个任务一个服务器）运行多个任务。客户端在不同的服务器上创建计算图，并通过调用远程会话在一台服务器上开始执行计算图。

作为示例，图 15-2 给出了连接到两个作业（名为 m1）的两个客户端。

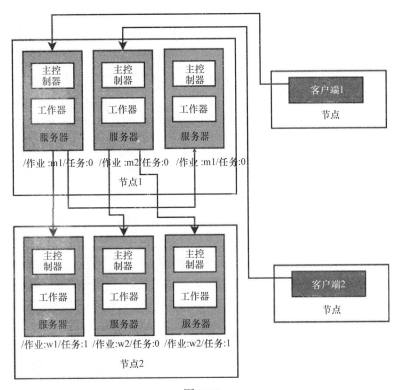

图 15-2

这两个节点分别运行三个任务，作业 w1 分布在两个节点上，而其他作业包含在一个节点中。

> TF 服务器实现为两个进程：主控制器（master）和工作器。主控制器协调其他任务的计算，工作器进行实际计算。这是在更高级别上，因此不必担心 TF 服务器的内部。这里是出于解释和示例的目的，因此仅涉及 TF 任务。

使用以下步骤创建和训练基于数据并行的模型：

1）定义集群规范；
2）创建用于控制任务的服务器；
3）定义要分配给参数服务器任务的变量节点；
4）定义要在所有工作器任务上复制的操作节点；
5）创建远程会话；
6）在远程会话中训练模型；
7）使用模型进行预测。

15.2.1 定义集群规范

要创建集群，首先要定义集群规范。集群规范通常包含两个作业：ps 用于创建参数服务器任务，worker 用于创建工作任务。worker 和 ps 作业包含物理节点列表，在这些节点上会运行各个任务。下面通过例子进行说明：

```
clusterSpec = tf.train.ClusterSpec({
    'ps': [
              'master0.neurasights.com:2222',    # /作业:ps/任务:0
              'master1.neurasights.com:2222'     # /作业:ps/任务:1
          ]
    'worker': [
              'worker0.neurasights.com:2222',    # /作业:worker/任务:0
              'worker1.neurasights.com:2222',    # /作业:worker/任务:1
              'worker0.neurasights.com:2223',    # /作业:worker/任务:2
              'worker1.neurasights.com:2223'     # /作业:worker/任务:3
          ]
})
```

此规范创建了两个作业，其中作业 ps 有两个任务，这两个任务分布在两个物理节点上；作业 worker 有 4 个任务，这些任务也分布在两个物理节点上。

示例代码会在 localhost 的不同端口上创建所有任务：

```
ps = [
        'localhost:9001',    # /作业:ps/任务:0
     ]
workers = [
        'localhost:9002',    # /作业:worker/任务:0
        'localhost:9003',    # /作业:worker/任务:1
        'localhost:9004',    # /作业:worker/任务:2
     ]
clusterSpec = tf.train.ClusterSpec({'ps': ps, 'worker': workers})
```

从上面的代码注释可以看出：任务的标识为 /job:<job name>/task:<task index>。

15.2.2 创建服务器实例

由于集群的每个任务包含一个服务器实例，因此在每个物理节点上，通过向服务器传递集群规范、自己的作业名称和任务索引启动服务器。服务器使用集群规范确定计算所涉及的其他节点。

```
server = tf.train.Server(clusterSpec, job_name="ps", task_index=0)
server = tf.train.Server(clusterSpec, job_name="worker", task_index=0)
server = tf.train.Server(clusterSpec, job_name="worker", task_index=1)
server = tf.train.Server(clusterSpec, job_name="worker", task_index=2)
```

在示例代码中，有一个 Python 文件，它将在所有物理机器上运行，该 python 文件的内容为：

```
server = tf.train.Server(clusterSpec,
                         job_name=FLAGS.job_name,
                         task_index=FLAGS.task_index,
                         config=config
                        )
```

在此代码中，job_name 和 task_index 是在命令行传递的参数。tf.flags 是一个解析器，可以访问命令行参数。在每个物理节点上按如下方式执行 Python 文件（如果您仅使用本地主机，则在同一节点上的不同的终端中执行）：

```
# 应使用适当的参数在每个物理节点中运行模型
$ python3 model.py --job_name='ps' --task_index=0
$ python3 model.py --job_name='worker' --task_index=0
$ python3 model.py --job_name='worker' --task_index=1
$ python3 model.py --job_name='worker' --task_index=2
```

> 为了在任何集群上运行代码，使其具有更高的灵活性，还可以通过命令行来传递要运行参数服务器和工作器程序的计算机列表：-ps='localhost:9001'-worker='localhost:9002, localhost:9003, localhost:9004'。然后解析这些值，并在集群规范字典中恰当地设置这些参数。

可通过配置对象确保参数服务器仅使用 CPU 而工作器任务使用 GPU：

```
config = tf.ConfigProto()
config.allow_soft_placement = True

if FLAGS.job_name=='ps':
    #输出(config.device_count['GPU'])
    config.device_count['GPU']=0
    server = tf.train.Server(clusterSpec,
                    job_name=FLAGS.job_name,
                    task_index=FLAGS.task_index,
                    config=config
                    )
    server.join()
    sys.exit('0')
elif FLAGS.job_name=='worker':
    config.gpu_options.per_process_gpu_memory_fraction = 0.2
    server = tf.train.Server(clusterSpec,
                    job_name=FLAGS.job_name,
                    task_index=FLAGS.task_index,
                    config=config
```

当工作器任务完成模型训练并退出后，参数服务器使用 server.join( ) 进行等待。图 15-3 为 GPU 在 4 台服务器上运行时的情形。

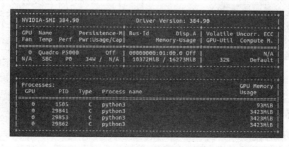

图 15-3

15.2.3 定义服务器和设备之间的参数和操作

可以使用在第 1 章中提到的 tf.device( ) 函数将参数放在 ps 任务上,将计算图的计算节点放在 worker 任务上。

 请注意,为了将计算图节点放置在特定设备上,还可以通过将设备字符串添加到任务字符串来实现,具体操作如下:/job:<jobname>/task:<task index>/device:<device type>:<device index>。

对于演示示例,将使用 TensorFlow 的 tf.train.replica_device_setter( ) 函数放置变量和操作。

1)首先,将工作器设备定义为当前工作器:

```
worker_device='/job:worker/task:{}'.format(FLAGS.task_index)
```

2)接下来,使用 replica_device_setter 定义设备函数,将集群规范和当前工作设备作为参数。通过 replica_device_setter 函数从集群规范中计算出参数服务器,如果有多个参数服务器,则默认以循环方式在这些服务器之间分配参数。在 tf.contrib 包中,可通过用户定义函数或预先构建的策略来修改参数设置策略。

```
device_func = tf.train.replica_device_setter(
    worker_device=worker_device,cluster=clusterSpec)
```

3)最后,在 tf.device(device_func) 块中创建计算图并训练。对于同步更新和异步更新,计算图的创建和训练是不同的,因此下面将分别介绍这些内容。

15.2.4 定义并训练计算图以进行异步更新

如图 15-4 所示(在前面也讨论过),在异步更新中,所有工作任务在准备就绪时发送参数更新,参数服务器更新参数并发回参数。参数更新没有同步(或等待和聚合)。

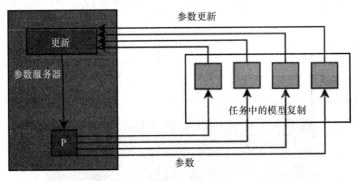

图 15-4

 此示例的完整代码在文件 ch-15_mnist_dist_async.py 中。建议基于自己的数据集修改和研究代码。

用以下步骤创建和训练计算图：
1）在 with 块中定义计算图：

```
with tf.device(device_func):
```

2）使用内置的 TensorFlow 函数创建全局步骤（step）变量：

```
global_step = tf.train.get_or_create_global_step()
```

3）这个变量也可以定义为：

```
tf.Variable(0,name='global_step',trainable=False)
```

4）按常规定义数据集、参数和超参数：

```
x_test = mnist.test.images
y_test = mnist.test.labels
n_outputs = 10   # 0~9 位数字
n_inputs = 784   # 总像素
learning_rate = 0.01
n_epochs = 50
batch_size = 100
n_batches = int(mnist.train.num_examples/batch_size)
n_epochs_print=10
```

5）按常规定义占位符、权重、偏差、对数、交叉熵、损失操作、训练操作、精确度：

```
# 输入图像
x_p = tf.placeholder(dtype=tf.float32,
                     name='x_p',
                     shape=[None, n_inputs])
# 目标输出
y_p = tf.placeholder(dtype=tf.float32,
                     name='y_p',
                     shape=[None, n_outputs])
w = tf.Variable(tf.random_normal([n_inputs, n_outputs],
                                  name='w'
                                  )
                )
b = tf.Variable(tf.random_normal([n_outputs],
                                  name='b'
                                  )
                )
logits = tf.matmul(x_p,w) + b

entropy_op = tf.nn.softmax_cross_entropy_with_logits(labels=y_p,
                                                     logits=logits
                                                     )
loss_op = tf.reduce_mean(entropy_op)

optimizer = tf.train.GradientDescentOptimizer(learning_rate)
train_op = optimizer.minimize(loss_op,global_step=global_step)

correct_pred = tf.equal(tf.argmax(logits, 1), tf.argmax(y_p, 1))
accuracy_op = tf.reduce_mean(tf.cast(correct_pred, tf.float32))
```

可通过改变这些定义来学习如何构建同步更新。

6）TensorFlow 提供了一个超类（supervisor class），可用它来创建训练会话，这在分布式训练设置中非常有用。下面的代码创建了一个 supervisor 对象：

```
init_op = tf.global_variables_initializer
sv = tf.train.Supervisor(is_chief=is_chief,
                    init_op = init_op(),
                    global_step=global_step)
```

7）使用 supervisor 对象创建会话，在此会话块中运行训练：

```
with sv.prepare_or_wait_for_session(server.target) as mts:
    lstep = 0

    for epoch in range(n_epochs):
        for batch in range(n_batches):
            x_batch, y_batch = mnist.train.next_batch(batch_size)
            feed_dict={x_p:x_batch,y_p:y_batch}
            _,loss,gstep=mts.run([train_op,loss_op,global_step],
                                feed_dict=feed_dict)
            lstep +=1
        if (epoch+1)%n_epochs_print==0:
            print('worker={},epoch={},global_step={}, \
            local_step={},loss={}'.
                format(FLAGS.task_index,epoch,gstep,lstep,loss))
    feed_dict={x_p:x_test,y_p:y_test}
    accuracy = mts.run(accuracy_op, feed_dict=feed_dict)
    print('worker={}, final accuracy = {}'
        .format(FLAGS.task_index,accuracy))
```

在启动参数服务器时，得到以下输出结果：

```
$ python3 ch-15_mnist_dist_async.py --job_name='ps' --task_index=0
I tensorflow/core/common_runtime/gpu/gpu_device.cc:1030] Found device 0 with properties:
    name: Quadro P5000 major: 6 minor: 1 memoryClockRate(GHz): 1.506
pciBusID: 0000:01:00.0
totalMemory: 15.89GiB freeMemory: 15.79 GiB
I tensorflow/core/common_runtime/gpu/gpu_device.cc:1120] Creating TensorFlow device (/device:GPU:0) -> (device: 0, name: Quadro P5000, pci bus id: 0000:01:00.0, compute capability:6.1
E1213 16:50:14.023235178   27224 ev_epoll1_linux.c:1051 grpc epoll fd: 23
I tensorflow/core/distributed_runtime/rpc/grpc_channel.cc:215] Initialize GrpcChannelCache for job ps -> {0 -> localhost:9001}
I tensorflow/core/distributed_runtime/rpc/grpc_channel.cc:215] Initialize GrpcChannelCache for job worker -> {0 -> localhost:9002, 1 -> localhost:9003,2 -> localhost:9004}
I tensorflow/core/distributed_runtime/rpc/grpc_server_lib.cc:324] **Started server with target: grpc://localhost:9001**
```

在启动工作器任务时，得到以下三个输出结果：

worker 1 的输出结果：

```
$ python3 ch-15_mnist_dist_async.py --job_name='worker'
--task_index=0
```
I tensorflow/core/common_runtime/gpu/gpu_device.cc:1030]
Found device 0 with properties:
  name: Quadro P5000 major: 6 minor: 1 memoryClockRate(GHz):
1.506 pciBusID: 0000:01:00.0
totalMemory: 15.89GiB freeMemory: 9.16GiB
I tensorflow/core/common_runtime/gpu/gpu_device.cc:1120] Creating
TensorFlow device (/device:GPU:0) -> (device: 0, name: Quadro P5000,
pci bus id: 0000:01:00.0, compute capability: 6.1)
E1213 16:50:37.516609689   27507 ev_epoll1_linux.c:1051]   grpc
epoll fd:23
I tensorflow/core/distributed_runtime/rpc/grpc_channel.cc:215]
Initialize GrpcChannelCache for job ps -> {0 -> localhost:9001}
I tensorflow/core/distributed_runtime/rpc/grpc_channel.cc:215]
Initialize GrpcChannelCache for job worker -> {0 -> localhost:9002,1->
localhost:9003, 2 -> localhost:9004}
I tensorflow/core/distributed_runtime/rpc/grpc_server_lib.cc:324]
Started server with target: grpc://localhost:9002
I tensorflow/core/distributed_runtime/master_session.cc:1004] Start
master session 1421824c3df413b5 with config: gpu_options {
per_process_gpu_memory_fraction: 0.2 } allow_soft_placement: true
worker=0,epoch=9,global_step=10896, local_step=5500, loss =
1.2575616836547852
worker=0,epoch=19,global_step=22453, local_step=11000, loss =
0.7158586382865906
worker=0,epoch=29,global_step=39019, local_step=16500, loss =
0.43712112307548523
worker=0,epoch=39,global_step=55513, local_step=22000, loss =
0.3935799300670624
worker=0,epoch=49,global_step=72002, local_step=27500, loss =
0.3877961337566376
worker=0, final accuracy = 0.8865000009536743

worker2 的输出结果：

```
$ python3 ch-15_mnist_dist_async.py --job_name='worker' --task_
index=1
```
I tensorflow/core/common_runtime/gpu/gpu_device.cc:1030] Found device
0 with properties:
  name: Quadro P5000 major: 6 minor: 1 memoryClockRate(GHz): 1.506
pciBusID: 0000:01:00.0
totalMemory: 15.89GiB freeMemory: 12.43GiB
I tensorflow/core/common_runtime/gpu/gpu_device.cc:1120] Creating
TensorFlow device (/device:GPU:0) -> (device: 0, name: Quadro P5000,
pci bus id: 0000:01:00.0, compute capability: 6.1)
E1213 16:50:36.684334877   27461 ev_epoll1_linux.c:1051]   grpc
epoll fd:23
I tensorflow/core/distributed_runtime/rpc/grpc_channel.cc:215]
Initialize GrpcChannelCache for job ps -> {0 -> localhost:9001}
I tensorflow/core/distributed_runtime/rpc/grpc_channel.cc:215]
Initialize GrpcChannelCache for job worker -> {0 -> localhost:9002,
1 -> localhost:9003, 2 -> localhost:9004}
I tensorflow/core/distributed_runtime/rpc/grpc_server_lib.cc:324]

```
Started server with target: grpc://localhost:9003
I tensorflow/core/distributed_runtime/master_session.cc:1004] Start
master session 2bd8a136213a1fce with config: gpu_options {
per_process_gpu_memory_fraction: 0.2 } allow_soft_placement: true
worker=1,epoch=9,global_step=11085, local_step=5500, loss =
0.6955764889717102
worker=1,epoch=19,global_step=22728, local_step=11000, loss =
0.5891970992088318
worker=1,epoch=29,global_step=39074, local_step=16500, loss =
0.4183048903942108
worker=1,epoch=39,global_step=55599, local_step=22000, loss =
0.32243454456329346
worker=1,epoch=49,global_step=72105, local_step=27500, loss =
0.5384714007377625
worker=1, final accuracy = 0.8866000175476074
```

worker3 的输出结果：

```
$ python3 ch-15_mnist_dist_async.py --job_name='worker' --task_index=2
I tensorflow/core/common_runtime/gpu/gpu_device.cc:1030] Found device
0 with properties:
    name: Quadro P5000 major: 6 minor: 1 memoryClockRate(GHz): 1.506
pciBusID: 0000:01:00.0
totalMemory: 15.89GiB freeMemory: 15.70GiB
I tensorflow/core/common_runtime/gpu/gpu_device.cc:1120] Creating
TensorFlow device (/device:GPU:0) -> (device: 0, name: Quadro P5000,
pci bus id: 0000:01:00.0, compute capability: 6.1)
E1213 16:50:35.568349791    27449 ev_epoll1_linux.c:1051]     grpc
epoll fd:23
I tensorflow/core/distributed_runtime/rpc/grpc_channel.cc:215]
Initialize GrpcChannelCache for job ps -> {0 -> localhost:9001}
I tensorflow/core/distributed_runtime/rpc/grpc_channel.cc:215]
Initialize GrpcChannelCache for job worker -> {0 -> localhost:9002, 1 ->
localhost:9003, 2 -> localhost:9004}
I tensorflow/core/distributed_runtime/rpc/grpc_server_lib.cc:324]Started
server with target: grpc://The full code for this example is in
ch-15_mnist_dist_sync.py. You are encouraged to modify and explore the
code with your own datasets.localhost:9004
I tensorflow/core/distributed_runtime/master_session.cc:1004] Start
master session cb0749c9f5fc163e with config: gpu_options {
per_process_gpu_memory_fraction: 0.2 } allow_soft_placement: true
I tensorflow/core/distributed_runtime/master_session.cc:1004] Start
master session 55bf9a2b9718a571 with config: gpu_options {
per_process_gpu_memory_fraction: 0.2 } allow_soft_placement: true
worker=2,epoch=9,global_step=37367, local_step=5500, loss =
0.8077645301818848
worker=2,epoch=19,global_step=53859, local_step=11000, loss =
0.26333487033843994
worker=2,epoch=29,global_step=70299, local_step=16500, loss =
0.6506651043891907
worker=2,epoch=39,global_step=76999, local_step=22000, loss =
0.20321622490882874
worker=2,epoch=49,global_step=82499, local_step=27500, loss =
0.4170967936515808
worker=2, final accuracy = 0.8894000053405762
```

这里输出了全局步骤和局部步骤，全局步骤表示所有工作器任务的步数，而局部步骤是该工作器任务中的计数，这就是为什么本地任务计数高达 27 500，而每个工作器的每个时期都相同，但是因为工作器正在按照自己的步调进行全局性的步骤，全局步骤的数量在工作器之间没有对称性。此外，可以看到每个工作器的最终精度会不同，因为每个工作器是在不同时间、不同参数的情况下得到的精度。

15.2.5 定义并训练计算图以进行同步更新

如图 15-5 所示（前面也介绍过），在同步更新中，任务将其更新发送到参数服务器，ps 任务等待接收所有更新并进行聚合，然后更新参数。工作任务在继续下一次计算参数更新之前等待更新。

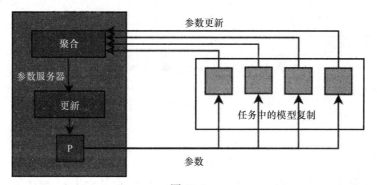

图 15-5

此示例的完整代码可在 Jupyter 笔记本 ch-15_mnist_dist_sync.py 中找到。建议读者基于自己的数据集修改和研究代码。

在同步更新时，需要对代码进行如下修改：

1）优化器需要包装在 SyncReplicaOptimizer 中。因此，在定义优化程序后，添加以下代码：

```
# SYNC: 添加下一行以使其同步更新
optimizer = tf.train.SyncReplicasOptimizer(optimizer,
    replicas_to_aggregate=len(workers),
    total_num_replicas=len(workers),
    )
```

2）像以前一样添加训练操作：

```
train_op = optimizer.minimize(loss_op,global_step=global_step)
```

3）接下来，添加初始化函数定义，具体的同步更新方法如下：

```
if is_chief:
    local_init_op = optimizer.chief_init_op()
else:
    local_init_op = optimizer.local_step_init_op()
chief_queue_runner = optimizer.get_chief_queue_runner()
init_token_op = optimizer.get_init_tokens_op()
```

4）可使用两个初始化函数创建不同的超对象（supervisor object）：

```
# SYNC: sv 针对同步更新进行了不同的初始化
sv = tf.train.Supervisor(is_chief=is_chief,
        init_op = tf.global_variables_initializer(),
        local_init_op = local_init_op,
        ready_for_local_init_op = optimizer.ready_for_local_init_op,
        global_step=global_step)
```

5）最后，在训练的会话块中，初始化同步变量，如果是主要的工作器任务，则启动队列运行器：

```
# SYNC: 如果添加了块以使其同步更新
if is_chief:
    mts.run(init_token_op)
    sv.start_queue_runners(mts, [chief_queue_runner])
```

其余代码与异步更新一样。

用于进行分布式训练的 TensorFlow 库和功能正在不断发展。因此，请注意添加新函数或函数名的变化。在撰写本书时所使用的是 TensorFlow 1.4。

15.3 总结

本章学习了如何使用 TensorFlow 集群在多台机器和设备上训练分布式模型，还基于 TensorFlow 代码介绍模型并行和数据并行策略。

在参数服务器中，参数更新可以按同步或异步更新的方式共享。本章介绍了如何实现同步和异步参数更新。借助本章中学到的技能，读者将能基于非常大的数据集构建和训练大模型。

下一章将学习如何在基于 iOS 和 Android 平台的移动和嵌入式设备上部署 TensorFlow 模型。

第 16 章 移动和嵌入式平台上的 TensorFlow 模型

TensorFlow 模型还可用于移动和嵌入式平台上的应用程序。TensorFlow Lite 和 TensorFlow Mobile 是针对资源受限的移动设备的两种 TensorFlow。TensorFlow Lite 的功能只是 TensorFlow Mobile 功能子集。由于较小的可执行程序和较少的依赖性，TensorFlow Lite 可以获得更好的性能。

如果要将 TensorFlow 集成到应用程序中，首先要使用本书介绍的技术训练模型并保存模型，然后在移动应用程序中使用保存的模型进行推理和预测。

为了使大家了解如何在移动设备上使用 TensorFlow 模型，本章将介绍以下主题：
- 移动平台上的 TensorFlow；
- Android 应用程序中的 TF Mobile；
- 演示 Android 上的 TF Mobile；
- 演示 iOS 上的 TF Mobile；
- TensorFlow Lite；
- 演示 Android 上的 TF Lite 应用程序；
- 演示 iOS 上的 TF Lite 应用程序。

16.1 移动平台上的 TensorFlow

在下面这些机器学习任务中都需要将 TensorFlow 集成到移动应用程序中：
- 语音识别；
- 图像识别；
- 手势识别；
- 光学字符识别；
- 图像或文本分类；
- 图像、文本或语音合成；
- 目标识别。

要在移动应用上运行 TensorFlow，主要涉及两方面的内容：
- 训练模型，保存可用于预测的模型；

- TensorFlow 可执行程序，它可以接收输入、运行模型、生成预测，并输出预测结果。高级架构如图 16-1 所示。

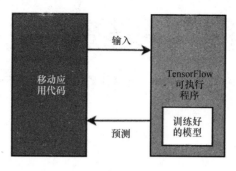

图 16-1

移动应用程序将输入传递给 TensorFlow 的可执行程序，该可执行程序使用训练的模型计算预测，并返回预测结果。

16.2　Android 应用程序中的 TF Mobile

在 Android 应用程序中，可使用 TensorFlow 生态系统中的接口类 TensorFlowInferenceInterface，并在 jar 文件 libandroid_tensorflow_inference_java.jar 中使用 TensorFlow Java API。可以使用 JCenter 中的 jar 文件从 ci.tensorflow.org 下载预编译的 jar，也可以自己构建。

为了让推理接口有效，需将其作为 JCenter 包；为了让推理接口包含在 Android 项目，需要将以下代码添加到 build.gradle 文件中：

```
allprojects {
    repositories {
        jcenter()
    }
}
dependencies {
    compile 'org.tensorflow:tensorflow-android:+'
}
```

若不使用 JCenter 中预先构建的二进制文件，可以按照以下链接的说明通过 Bazel 或 Cmake 自行构建：https://github.com/tensorflow/tensorflow/blob/r1.4/tensorflow/contrib/android/README.md。

在 Android 项目中配置 TF 库后，可以通过以下 4 个步骤调用 TF 模型：

1）加载模型：

```
TensorFlowInferenceInterface inferenceInterface =
    new TensorFlowInferenceInterface(assetManager, modelFilename);
```

2）将输入数据传递给 TensorFlow 的可执行程序：

```
inferenceInterface.feed(inputName,
    floatValues, 1, inputSize, inputSize, 3);
```

3）运行预测或推理：

    inferenceInterface.run(outputNames, logStats);

4）接收 TensorFlow 可执行程序的输出：

    inferenceInterface.fetch(outputName, outputs);

16.3 演示 Android 上的 TF Mobile

本节将学习如何重新创建 TensorFlow 团队在其官方资料库中用于演示的 Android 应用程序。这个演示程序需要在 Android 设备上安装以下 4 个应用：

- TF Classify：这是一个对象识别应用程序，用于识别设备摄像头输入中的图像，并通过预定义类别对其进行分类。该应用程序不会学习新类型的图片，但会尝试按已有的类别对新图片分类。该应用程序使用 Google 预先训练好的 inception 模型进行构建。
- TF Detect：这是一个目标检测应用程序，可检测图像中的多个目标。在摄像头不断输入图像的情况下，它会一直识别目标。
- TF Stylize：这是一个样式迁移应用程序，可将选定的预定义样式作为设备摄像头的输入。
- TF Speech：这是一个语音识别应用程序，用于识别语音，如果与应用程序中的某个预定义命令匹配，则会在设备屏幕上突出显示该命令。

> 该演示程序仅适用于 API 级别大于 21 的 Android 设备，并且该设备必须具有支持 FOCUS_MODE_CONTINUOUS_PICTURE 的摄像头。如果设备摄像头不能支持此功能，则必须添加作者提交给 TensorFlow 的路径：https://github.com/tensorflow/tensorflow/pull/15489/files。

在设备上构建和部署该演示应用程序的最简单方法是使用 Android Studio。若要以这种方式构建，请按照下列步骤操作：

1）安装 Android Studio。通过以下链接的说明在 Ubuntu 16.04 上安装 Android Studio：https://developer.android.com/studio/install.html。

2）查看 TensorFlow 资料库，并使用上面那个注意事项⊖中所提到的补丁。假设读者已经导出 tensorflow 文件夹中的代码。

3）使用 Android Studio 在路径中打开 Android 项目 ~/ tensorflow/ tensorflow/examples/Android。得到的结果与图 16-2 类似。

4）从图 16-2 的左侧栏中展开 Gradle Scripts 选项，然后打开 build.gradle 文件。

5）在 build.gradle 文件中找到 def nativeBuildSystem 定义并将其设置为 "none"。在导出的代码中，下面的定义位于第 43 行：

    def nativeBuildSystem = 'none'

⊖ 在这里下载补丁：https://github.com/ tensorflow/tensorflow/pull/15489/files。

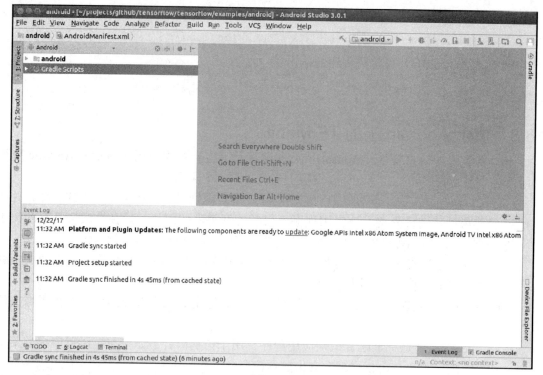

图 16-2

6）构建演示并在真实或模拟设备上运行。本书在这些设备上测试了应用（见图 16-3）。

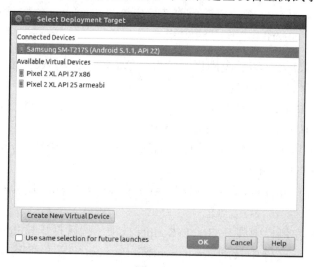

图 16-3

7）还可以构建 apk，并在虚拟或实际连接的设备上安装 apk 文件。一旦应用程序安装在设备上，将看到之前讨论过的 4 个应用程序。在 Android 仿真器中的 TF 应用程序示意图如图 16-4 所示。

第 16 章 | 移动和嵌入式平台上的 TensorFlow 模型

图 16-4

 还可以按照以下链接中的说明通过 Bazel 或 Cmake 构建整个演示应用程序：
https://github.com/tensorflow/tensorflow/tree/r1.4/tensorflow/examples/android。

16.4　iOS 应用程序中的 TF Mobile

TensorFlow 通过以下步骤支持 iOS 应用：

1）通过在项目的根目录中添加名为 Profile 的文件，在应用程序中包含 TF Mobile。将以下内容添加到配置文件：

```
target 'Name-Of-Your-Project'
        pod 'TensorFlow-experimental'
```

2）运行 pod install 命令下载并安装 TensorFlow 的实验 pod。

3）运行 myproject.xcworkspace 命令打开工作区，以便将预测代码添加到应用程序逻辑中。

> 为 iOS 项目创建自己的 TensorFlow 可执行程序，并按照以下链接中的说明操作：https://github.com/tensorflow/tensorflow/ tree/master/tensorflow/examples/ios。

在 iOS 项目中配置 TF 库后，可以通过以下 4 个步骤调用 TF 模型：

1）加载模型：

```
PortableReadFileToProto(file_path, &tensorflow_graph);
```

2）创建对话：

```
tensorflow::Status s = session->Create(tensorflow_graph);
```

3）运行预测或推理并获得输出结果：

```
std::string input_layer = "input";
std::string output_layer = "output";
std::vector<tensorflow::Tensor> outputs;
tensorflow::Status run_status = session->Run(
    {{input_layer, image_tensor}},
    {output_layer}, {}, &outputs);
```

4）获取输出数据：

```
tensorflow::Tensor* output = &outputs[0];
```

16.5　演示 iOS 上的 TF Mobile

要在 iOS 上构建所要演示的应用程序，需要 Xcode 7.3 或更高版本。可按照以下步骤构建基于 iOS 的应用程序：

1）查看 tensorflow 文件夹中的 TensorFlow 代码。

2）打开终端窗口，并从主文件夹执行以下命令下载 Inception V1 模型，提取标签和图形文件，并将这些文件移动到应用程序代码的数据文件夹中：

```
$ mkdir -p ~/Downloads
$ curl -o ~/Downloads/inception5h.zip \
https://storage.googleapis.com/download.tensorflow.org/models/inception5h.zip \
  && unzip ~/Downloads/inception5h.zip -d ~/Downloads/inception5h
$ cp ~/Downloads/inception5h/* \
  ~/tensorflow/tensorflow/examples/ios/benchmark/data/
$ cp ~/Downloads/inception5h/* \
```

```
~/tensorflow/tensorflow/examples/ios/camera/data/
$ cp ~/Downloads/inception5h/* \
    ~/tensorflow/tensorflow/examples/ios/simple/data/
```

3）转到其中一个示例文件夹并下载实验 pod：

```
$ cd ~/tensorflow/tensorflow/examples/ios/camera
$ pod install
```

4）打开 Xcode 的工作区：

```
$ open tf_simple_example.xcworkspace
```

5）在设备模拟器中运行示例应用程序。示例应用程序将显示"运行模型"按钮。相机应用程序需要连接 Apple 设备，而其他两个应用也可以在模拟器中运行。

16.6 TensorFlow Lite

在编写本书时，TensorFlow Lite（简称 TF Lite）才推出来，并且仍在开发人员视图中。TF Lite 在功能上是 TensorFlow Mobile 和 TensorFlow 的一个非常小的子集，因此使用 TF Lite 生成的可执行程序非常小，而且性能非常好。除了减小了可执行程序的大小外，TensorFlow 还采用了其他技术，例如：

- 针对各种设备和移动架构优化了内核；
- 量化计算中使用的值；
- 激活函数是预融合的；
- 会利用设备上专用机器学习软件或硬件，例如 Android NN API。

在 TF Lite 中使用模型的工作流程如下：

1）**获取模型**：可以训练自己的模型或选择其他的预训练模型，可直接使用预训练模型，也可使用自己的数据重新训练模型，也可以在修改模型的某些部分后重新训练。只要在文件中使用扩展名为 .pb 或 .pbtxt 的训练模型，就可以继续执行下一步。利用前面章节介绍的方法保存模型。

2）**检查点模型**：模型文件只包含计算图的结构，所以需要保存检查点文件。检查点文件包含序列化的文件、模型变量，例如权重和偏差。可用前几章介绍的方法保存检查点。

3）**冻结模型**：合并检查点和模型文件，也称为冻结图。TensorFlow 为此步骤提供了 freeze_graph 工具，可以按如下方式冻结模型：

```
$ freeze_graph
    --input_graph=mymodel.pb
    --input_checkpoint=mycheckpoint.ckpt
    --input_binary=true
    --output_graph=frozen_model.pb
    --output_node_name=mymodel_nodes
```

4）**转换模型**：使用 TensorFlow 提供的 toco 工具可将步骤 3 中的冻结模型转换为 TF Lite 格式：

```
$ toco
    --input_file=frozen_model.pb
    --input_format=TENSORFLOW_GRAPHDEF
    --output_format=TFLITE
    --input_type=FLOAT
    --input_arrays=input_nodes
    --output_arrays=mymodel_nodes
    --input_shapes=n,h,w,c
```

5）在步骤 4 中保存 .tflite 模型可将该模型用于 Android 或 iOS 应用程序中，这些应用程序会使用 TF Lite 可执行程序进行推理。在应用程序中包含 TF Lite 可执行程序的方式还在不断改进，因此建议读者按照以下链接中的信息在自己的 Android 或 iOS 应用程序中包含 TF Lite 可执行程序：https://github.com/tensorflow/tensorflow/tree/master/tensorflow/contrib/lite/g3doc。

通常，可以使用 graph_transforms：summarize_graph 工具修改在步骤 1 中获得的模型。修改后的模型将只有在推理或预测时从输入到输出的路径。删除只在训练或调试才需要的任何节点和路径（例如保存检查点），从而使最终的模型变得非常小。

官方的 TensorFlow 资料库提供了一个基于 TF Lite 的演示程序，该程序使用预训练的 mobilenet 对摄像头输入的图像进行分类，其分类的类别数为 1001 个。演示应用程序会显示前三个类别的概率。

16.7　演示 Android 上的 TF Lite 应用程序

要在 Android 上构建 TF Lite 应用程序，请按照下列步骤操作：

1）安装 Android Studio。通过以下链接的说明在 Ubuntu 16.04 上安装 Android Studio：https://developer.android.com/studio/install.html。

2）查看 TensorFlow 资料库，并应用前面介绍的补丁。假设已经导出了 tensorflow 文件夹中的代码。

3）使用 Android Studio 在路径中打开 Android 项目 ~/tensorflow/tensorflow/contrib/lite/java/demo。如果报出缺少 SDK 或 Gradle 组件的错误，请安装这些组件并同步 Gradle。

4）构建项目并在 API > 21 的虚拟设备上运行。

虽然会收到以下警告，但这已经构建成功了。如果构建失败，可能需要解决警告提示的问题：

```
Warning:The Jack toolchain is deprecated and will not
run. To enable support for Java 8 language features built
into the plugin, remove 'jackOptions { ... }' from your
build.gradle file, and add

android.compileOptions.sourceCompatibility 1.8
android.compileOptions.targetCompatibility 1.8
```

```
Future versions of the plugin will not support usage of
'jackOptions' in build.gradle.
To learn more, go to
https://d.android.com/r/tools/java-8-support-message.html

Warning:The specified Android SDK Build Tools version
(26.0.1) is ignored, as it is below the minimum supported
version (26.0.2) for Android Gradle Plugin 3.0.1.
Android SDK Build Tools 26.0.2 will be used.
To suppress this warning, remove "buildToolsVersion
'26.0.1'" from your build.gradle file, as each version of
the Android Gradle Plugin now has a default version of
the build tools.
```

也可以使用 Bazel 从源代码构建整个演示程序，构建过程可参考以下链接：https://github.com/tensorflow/tensorflow/tree/master/ tensorflow/contrib/lite。

16.8　演示 iOS 上的 TF Lite 应用程序

要在 iOS 上构建要演示的应用程序，需要 Xcode 7.3 或更高版本。按照以下步骤进行构建：

1）查看 tensorflow 文件夹中的 TensorFlow 代码。

2）根据以下链接的说明构建适用于 iOS 的 TF Lite 可执行程序：https: // github.com/tensorflow/tensorflow/tree/master/tensorflow/contrib/lite。

3）导航到示例文件夹并下载 pod：

```
$ cd ~/tensorflow/tensorflow/contrib/lite/examples/ios/camera
$ pod install
```

4）打开 Xcode 工作区：

```
$ open tflite_camera_example.xcworkspace
```

5）在设备模拟器中运行示例应用程序。

16.9　总结

本章学习了如何在移动应用程序和设备上使用 TensorFlow 模型。TensorFlow 提供了两种在移动设备上运行的开发包：TF Mobile 和 TF Lite。本章还学习了如何为 iOS 和 Android 构建 TF Mobile 和 TF Lite 应用程序。本章使用了 TensorFlow 演示应用程序作为示例，鼓励读者研究这些演示应用程序的源代码，为了支持具体的移动应用，可使用 TF Mobile 和 TF Lite 构建机器学习模型。

下一章将学习如何在统计软件 R 中使用 TensorFlow，这会基于 RStudio 所发布的 R 包上进行介绍。

第 17 章 R 中的 TensorFlow 和 Keras

R 是一个开源平台,它是一种统计计算的环境和语言。R 还有一个桌面 IDE 和一个 Web 的 IDE,它们统称为 R Studio。有关 R 的更多信息,访问以下链接:http://www.r-project.org/。R 通过提供以下 R 包实现对 TensorFlow 和 Keras 的支持:
- tensorflow 包提供对 TF 核心 API 的支持;
- tfestimators 包提供对 TF 估计器 API 的支持;
- keras 包提供对 Keras API 的支持;
- tfruns 包用于可视化 TensorBoard 模型和训练过程。

本章将学习如何在 R 中使用 TensorFlow,这将涵盖以下主题:
- 在 R 中安装 TensorFlow 和 Keras 软件包;
- R 中的 TF 核心 API;
- R 中的 TF Estimator API;
- R 中的 Keras API;
- R 中的 TensorBoard;
- R 中的 tfruns 包。

17.1 在 R 中安装 TensorFlow 和 Keras 软件包

要在 R 中安装支持 TensorFlow 和 Keras 的三个 R 软件包,请在 R 中执行以下命令。

1)首先,安装 devtools:

```
install.packages("devtools")
```

2)安装 tensorflow 和 tfestimators 包:

```
devtools::install_github("rstudio/tensorflow")
devtools::install_github("rstudio/tfestimators")
```

3)加载 tensorflow 库并安装所需的函数:

```
library(tensorflow)
install_tensorflow()
```

4)默认情况下,安装函数会创建虚拟环境并在虚拟环境中安装 tensorflow 包。
可以使用 method 参数指定 4 种有效的安装方法,见表 17-1。

表 17-1

auto	自动选择当前平台的默认值
virtualenv	安装到位于 ~/.virtualenvs/r-tensorflow 的虚拟环境中
conda	安装到名为 r-tensorflow 的 Anaconda Python 环境中
system	安装到系统的 Python 环境中

5）默认情况下，安装函数只会安装 TensorFlow 的 CPU 版本。要安装 GPU 版本，请使用版本参数，见表 17-2。

表 17-2

GPU	安装 tensorflow-gpu
nightly	以 nightly 方式安装 CPU 版本
nightly-gpu	以 nightly 方式安装 GPU 版本
n.n.n	安装指定版本，例如 1.3.0
n.n.n-gpu	安装指定的 GPU 版本，例如 1.3.0

如果希望 TensorFlow 库使用具体版本的 Python，请使用以下函数或设置 TENSOR-FLOW_PYTHON 环境变量：

- use_python('/usr/bin/python2')
- use_virtualenv('~/venv')
- use_condaenv('conda-env')
- Sys.setenv(TENSORFLOW_PYTHON='/usr/bin/python2')

使用以下命令在 Ubuntu 16.04 上的 R 中安装 TensorFLow：
install_tensorflow(version="gpu")
请注意，在编写本书时，安装不支持 Python 3。

6）安装 keras 包：

devtools::install_github("rstudio/keras")

7）在虚拟环境中安装 Keras：

library(keras)
install_keras()

8）要安装 GPU 版本，请使用：

install_keras(tensorflow = "gpu")

9）安装 tfruns 包：

devtools::install_github("rstudio/tfruns")

17.2 R 中的 TF 核心 API

在第 1 章中介绍了 TensorFlow 的核心 API。在 R 中，这些 API 由 tensorflow R 包实现。下面将演示一个基于 MLP 模型的例子，它能对 MNIST 数据集的手写数字进行分类，可以从以下链接下载相关代码：https://tensorflow.rstudio.com/ tensorflow / articles / examples / mnist_softmax.html。

 可以按照 Jupyter R 笔记本 ch-17a_TFCore_in_R 中的代码进行操作。

1）首先，加载库：

```
library(tensorflow)
```

2）定义超参数：

```
batch_size <- 128
num_classes <- 10
steps <- 1000
```

3）准备数据：

```
datasets <- tf$contrib$learn$datasets
mnist <- datasets$mnist$read_data_sets("MNIST-data", one_hot = TRUE)
```

数据用 TensorFlow 数据集库加载，而且已经归一化到 [0,1] 范围。

4）定义模型：

```
# 创建模型
x <- tf$placeholder(tf$float32, shape(NULL, 784L))
W <- tf$Variable(tf$zeros(shape(784L, num_classes)))
b <- tf$Variable(tf$zeros(shape(num_classes)))
y <- tf$nn$softmax(tf$matmul(x, W) + b)

# 定义损失和优化器
y_ <- tf$placeholder(tf$float32, shape(NULL, num_classes))
cross_entropy <- tf$reduce_mean(-tf$reduce_sum(y_ * log(y), reduction_indices=1L))
train_step <-
tf$train$GradientDescentOptimizer(0.5)$minimize(cross_entropy)
```

5）训练模型：

```
# 创建会话并初始化变量
sess <- tf$Session()
sess$run(tf$global_variables_initializer())

# 训练
for (i in 1:steps) {
  batches <- mnist$train$next_batch(batch_size)
  batch_xs <- batches[[1]]
  batch_ys <- batches[[2]]
```

```
        sess$run(train_step,
                feed_dict = dict(x = batch_xs, y_ = batch_ys))
    }
```

6)评估模型:

```
    correct_prediction <- tf$equal(tf$argmax(y, 1L), tf$argmax(y_, 1L))
    accuracy <- tf$reduce_mean(tf$cast(correct_prediction, tf$float32))
    score <-sess$run(accuracy,
                feed_dict = dict(x = mnist$test$images,
                                 y_ = mnist$test$labels))

    cat('Test accuracy:', score, '\n')
```

输出结果如下:

```
    Test accuracy: 0.9185
```

太棒了!

 为了在 R 中得到更多基于 TF Core 的示例, 可访问以下链接: https://tensorflow.rstudio.com/tensorflow/articles/examples/。查看 tensorflow R 软件包的更多文档可访问以下链接: https://tensorflow.rstudio.com/tensorflow/reference/。

17.3 R 中的 TF Estimator API

第 2 章介绍了 TensorFlow Estimator API。在 R 中,这些 API 用 tfestimator R 包实现。

下面将演示一个基于 MLP 模型的例子,它能对 MNIST 数据集的手写数字进行分类。可以从以下链接下载相关代码: https://tensorflow.rstudio.com/tfestimators/articles/examples/mnist.html。

可以在 Jupyter R 笔记本 ch-17b_TFEstimator_in_R 中找相关的代码。

1)首先,加载库:

```
    library(tensorflow)
    library(tfestimators)
```

2)定义超参数:

```
    batch_size <- 128
    n_classes <- 10
    n_steps <- 100
```

3)准备数据

```
    # 初始化数据目录
    data_dir <- "~/datasets/mnist"
    dir.create(data_dir, recursive = TRUE, showWarnings = FALSE)

    # 下载 MNIST 数据集,并将它们读入 R 中
    sources <- list(
      train = list(
        x =
```

```r
      "https://storage.googleapis.com/cvdf-datasets/mnist/train-images-id
x3-ubyte.gz",
      y =
"https://storage.googleapis.com/cvdf-datasets/mnist/train-labels-id
x1-ubyte.gz"
    ),
    test = list(
      x =
"https://storage.googleapis.com/cvdf-datasets/mnist/t10k-images-idx
3-ubyte.gz",
      y =
"https://storage.googleapis.com/cvdf-datasets/mnist/t10k-labels-idx
1-ubyte.gz"
    )
)

# 读取MNIST文件（以IDX格式编码）
read_idx <- function(file) {
    #创建文件的二进制连接
    conn <- gzfile(file, open = "rb")
    on.exit(close(conn), add = TRUE)
    # 将魔数读取为4个字节的序列
    magic <- readBin(conn, what="raw", n=4, endian="big")
    ndims <- as.integer(magic[[4]])
    # 读取大小（32位整数）
    dims <- readBin(conn,what="integer",n=ndims,endian="big")
    # 以原始向量的形式读取剩余部分
    data <- readBin(conn,what="raw",n=prod(dims),endian="big")
    # 转换为整数向量
    converted <- as.integer(data)
    # 返回一维数组的普通向量
    if (length(dims) == 1)
        return(converted)
    # 将三维数据包装成矩阵
    matrix(converted,nrow=dims[1],ncol=prod(dims[-1]),byrow=TRUE)
}

mnist <- rapply(sources,classes="character",how
="list",function(url) {
    # 下载并从URL中提取文件
    target <- file.path(data_dir, basename(url))
    if (!file.exists(target))
        download.file(url, target)
    # 读取IDX文件
    read_idx(target)
})

# 将训练数据强度转换为0~1范围
mnist$train$x <- mnist$train$x / 255
mnist$test$x <- mnist$test$x / 255
```

从下载的 gzip 文件中读取数据，然后归一化到 [0,1] 范围。

4）定义模型

```
# 构建线性分类器
classifier <- linear_classifier(
  feature_columns = feature_columns(
    column_numeric("x", shape = shape(784L))
  ),
  n_classes = n_classes # 10 位数字
)

# 构建输入函数发生器
mnist_input_fn <- function(data, ...) {
  input_fn(
    data,
    response = "y",
    features = "x",
    batch_size = batch_size,
    ...
  )
}
```

5）训练模型

```
train(classifier,input_fn=mnist_input_fn(mnist$train),steps=n_steps
)
```

6）评估模型

```
evaluate(classifier,input_fn=mnist_input_fn(mnist$test),steps=200)
```

输出的结果如下：

```
Evaluation completed after 79 steps but 200 steps was specified
```

平均损失	损失		全局步骤精度
0.35656	45.13418	100	0.9057

太棒了！

为了在 R 中得到更多基于 TF Estimator 的示例，可访问以下链接：https://tensorflow.rstudio.com/tfestimators/articles/examples/。有关 tensorflow R 包的更多文档，请访问以下链接：https://tensorflow.rstudio.com/tfestimators/reference/。

17.4　R 中的 Keras API

在第 3 章中介绍了 Keras API。在 R 中，这些 API 在 keras R 包中被实现。keras R 包实现了 Keras Python 接口的大部分功能，包括序列化 API 和功能性 API。

例如，下面将演示一个基于 MLP 模型的例子，它能对 MNIST 数据集的手写数字进行分类。可从以下链接下载相关代码：https://keras.rstudio.com/ articles / examples / mnist_mlp.html。

读者可以按照 Jupyter R 笔记本 ch-17c_Keras_in_R 中的代码进行操作。

1）首先，加载库：

```
library(keras)
```

2）定义超参数：

```
batch_size <- 128
num_classes <- 10
epochs <- 30
```

3）准备数据：

```
# 训练集和测试集之间的数据，混合和分离
c(c(x_train, y_train), c(x_test, y_test)) %<-% dataset_mnist()

x_train <- array_reshape(x_train, c(nrow(x_train), 784))
x_test <- array_reshape(x_test, c(nrow(x_test), 784))

# 将RGB值转换为[0,1]范围
x_train <- x_train / 255
x_test <- x_test / 255

cat(nrow(x_train), 'train samples\n')
cat(nrow(x_test), 'test samples\n')

# 将类向量转换为二值矩阵
y_train <- to_categorical(y_train, num_classes)
y_test <- to_categorical(y_test, num_classes)
```

从注释可知：数据从 Keras 数据集库加载，然后转换为二维数据，并归一化到 [0,1] 范围。

4）定义模型：

```
model <- keras_model_sequential()
model %>%
  layer_dense(units=256,activation='relu',input_shape=c(784)) %>%
  layer_dropout(rate = 0.4) %>%
  layer_dense(units = 128, activation = 'relu') %>%
  layer_dropout(rate = 0.3) %>%
  layer_dense(units = 10, activation = 'softmax')

summary(model)

model %>% compile(
  loss = 'categorical_crossentropy',
  optimizer = optimizer_rmsprop(),
  metrics = c('accuracy')
)
```

5）定义和编译序列模型，得到如下的模型定义：

```
Layer (type)                  Output Shape              Param #
================================================================
dense_26 (Dense)              (None, 256)               200960

dropout_14 (Dropout)          (None, 256)               0

dense_27 (Dense)              (None, 128)               32896

dropout_15 (Dropout)          (None, 128)               0

dense_28 (Dense)              (None, 10)                1290
================================================================
Total params: 235,146
Trainable params: 235,146
Non-trainable params: 0
```

6）训练模型：

```
history <- model %>% fit(
    x_train, y_train,
    batch_size = batch_size,
    epochs = epochs,
    verbose = 1,
    validation_split = 0.2
)

plot(history)
```

fit 函数的输出存储在 history 对象中，该对象包含来自训练时期的损失值和度量值。绘制 history 对象中的数据，结果如图 17-1 所示。

7）评估模型：

```
score <- model %>% evaluate(
    x_test, y_test,
    verbose = 0
)

# 输出指标
cat('Test loss:', score[[1]], '\n')
cat('Test accuracy:', score[[2]], '\n')
```

输出结果如下：

```
Test loss: 0.1128517
Test accuracy: 0.9816
```

太棒了！

> 在下面的链接中能找到基于 R 的 Keras 的更多示例：https://keras.rstudio.com/articles/examples/index.html。有关 Keras R 包的更多文档可在以下链接中找到：https://keras.rstudio.com/reference/index.htm。

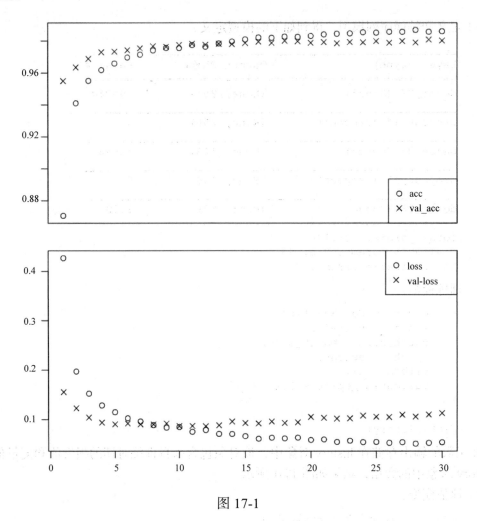

图 17-1

17.5　R 中的 TensorBoard

　可以在 Jupyter R 笔记本 ch-17d_TensorBoard_in_R 中找到相关的代码。

可以使用 tensorboard( ) 函数查看 TensorBoard，具体操作如下：

```
tensorboard('logs')
```

其中，'logs' 是将要创建 TensorBoard 日志的文件夹。

数据将被记录并按执行周期显示。在 R 中，收集 TensorBoard 的数据取决于所使用的包：

- 如果使用 tensorflow 包，请对计算图添加 tf $ summary $ scalar 操作；

- 如果使用 tfestimators 包，那么 TensorBoard 数据会自动写入创建估算器时 model_dir 参数所指向的路径；
- 如果使用 keras 包，在使用 fit( ) 函数训练模型时必须包含 callback_tensorboard( ) 函数。修改之前 Keras 示例中的训练，具体代码如下：

```
# 训练模型 --------
tensorboard("logs")

history <- model %>% fit(
    x_train, y_train,
    batch_size = batch_size,
    epochs = epochs,
    verbose = 1,
    validation_split = 0.2,
    callbacks = callback_tensorboard("logs")
)
```

当执行 Jupyter R 笔记本时，训练单元将获得如下的结果：

```
Started TensorBoard at http://127.0.0.1:4233
```

当点击链接时，会看到在 TensorBoard 中绘制的标量，如果 17-2 所示。

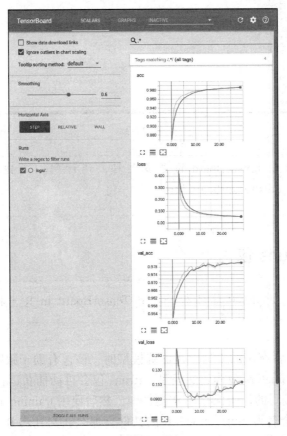

图 17-2

单击 Graphs 选项卡，在 TensorBoard 中会看到如图 17-3 所示的计算图。

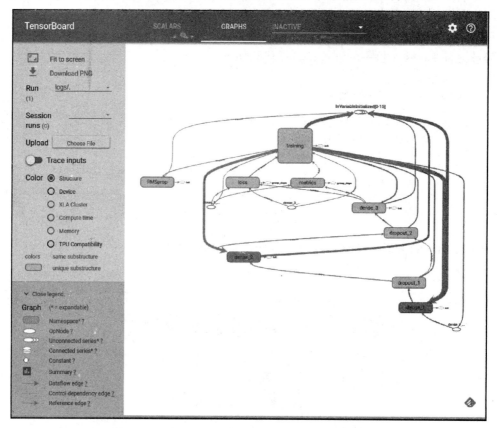

图 17-3

有关 R 中 TensorBoard 的更多文档，请访问以下链接：https://tensorflow.rstudio.com/tools/tensorboard.html。

17.6　R 中的 tfruns 包

可以在 Jupyter R 笔记本 ch-17d_TensorBoard_in_R 中找到相关的代码进行操作。

tfruns 包提供了非常有用的工具，在训练模型时，该包有助于跟踪多次运行。在 R 中，对于使用 keras 和 tfestimators 包构建的模型，tfruns 包会自动捕获运行数据。使用 tfruns 非常简单容易。只需将编写的代码保存在 R 文件中，然后使用 training_run( ) 函数执行这个文件。例如，如果有一个名为 mnist_model.R 的文件，那么在交互式 R 控制台中可通过 training_run( ) 函数来执行该文件，具体的代码如下：

```
library(tfruns)
training_run('mnist_model.R')
```

训练完成后，将自动显示摘要窗口。图 17-4 是运行 mnist_mlp.R 得到的结果，这个 R 文件是从 tfruns GitHub 资料库 (https://github.com/rstudio/tfruns/blob/master/ inst/examples/ mnist_mlp/mnist_mlp.R) 中获得的。

图 17-4

在"Viewer"窗口中，输出选项卡包含图 17-5 所示图形。

tfruns 软件包会在 RStudio 中安装一个插件，可以从 Addins 菜单中访问。该软件包还允许用户比较多个运行结果，并且可以将运行报告发布到 RPubs 或 RStudio Connect 上，用户也可以选择在本地保存报告。

> **TIP** 有关 tfruns 软件包的更多文档，请访问以下链接：
> https://tensorflow.rstudio.com/tools/tfruns/reference/；
> https://tensorflow.rstudio.com/tools/tfruns/articles/overview.html。

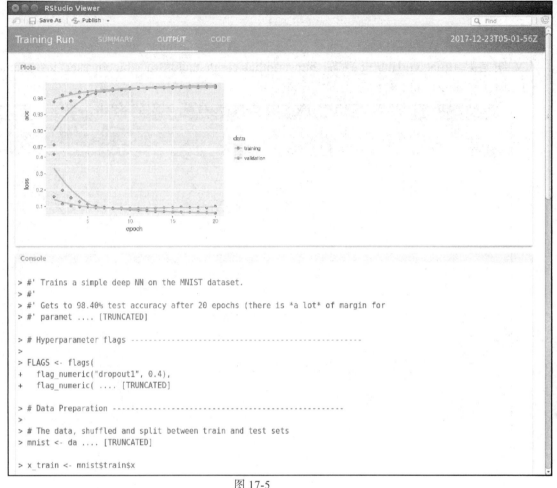

图 17-5

17.7 总结

本章介绍了如何在 R 中使用 TensorFlow Core、TensorFlow Estimator 和 Keras 包构建和训练机器学习模型。本章在 RStudio 中演示了 MNIST 示例，并提供进一步学习 TensorFlow 和 Keras R 包的文档的链接，还介绍如何使用来自 R 内部的可视化工具 TensorBoard。本章也介绍了一个来自 R Studio 的新工具——fruns，该工具允许用户为多个执行过程创建、分析和比较报告，并在本地保存或发布这些报告。

具有直接在 R 中工作的能力很有用，因为大量的数据科学和机器学习代码是使用 R 编写的，现在可以将 TensorFlow 集成到代码库中并在 R 环境中运行。

下一章将介绍如何调试构建和训练 TensorFlow 模型的代码。

第 18 章
调试 TensorFlow 模型

本书前面介绍了如何用 TensorFlow 构建和训练用于各种任务的预测模型。在训练模型时，可以构建计算图、运行计算图来进行训练，并通过评估计算图来进行预测。重复这些任务，直到获得满意的模型，然后保存计算图与学习的参数。在生产环境中，可构建计算图，也可从文件中恢复，并使用参数填充。

构建深度学习模型是一项复杂的技术，TensorFlow API 及其生态系统也很复杂。当在 TensorFlow 中构建和训练模型时，有时会得到各种的错误，或者模型不能按预期工作。例如，人们经常碰到以下问题：
- 在损失函数和度量的输出中有 NaN；
- 即使经过多次迭代，也没有得到改善。

在出现这些情形时，需要调试基于 TensorFlow API 编写的代码。

为了修复代码以使其能正常工作，可以使用调试器或平台提供的其他方法和工具，例如 Python 中的 Python 调试器（pdb）和 Linux OS 中的 GNU 调试器（gdb）。当出现问题时，TensorFlow API 还提供了一些额外的支持以修复代码。

本章将学习 TensorFlow 中可用的工具和技术以帮助调试：
- 使用 tf.Session.run( ) 获取张量值；
- 使用 tf.Print( ) 输出张量值；
- 使用 tf.Assert( ) 断言条件；
- 使用 TensorFlow 调试器（tfdbg）进行调试。

18.1 使用 tf.Session.run( ) 获取张量值

可以使用 tf.Session.run( ) 获取要输出的张量值。这些值作为 NumPy 数组返回，可以使用 Python 语句打印出来。这是最简单、最便捷的方法，最大的缺点是计算图会执行所有依赖路径，从获取的张量开始，如果这些路径包括训练操作，则会增加一步或一个周期。

因此，大多数情况下不会调用 tf.Session.run( ) 获取计算图中间的张量，但也可以执行整个计算图并获取所有张量，有些张量需要调试，有些不需要。

 函数 tf.Session.partial_run( ) 也适用于执行部分计算图的情形，但是该函数是一个高度实验性的 API，尚未准备正式推出。

18.2 使用 tf.Print( ) 输出张量值

可用 tf.Print( ) 输出相关信息。可将一个张量传递给 tf.Print( )，以便当包含 tf.Print( ) 节点的路径被执行时，能在错误控制台中输出相应的值。tf.Print( ) 函数的定义如下：

```
tf.Print(
    input_,
    data,
    message=None,
    first_n=None,
    summarize=None,
    name=None
)
```

该函数的各个参数的含义如下：
- input_ 是一个从函数返回的张量，没有任何操作。
- data 是要输出的张量列表。
- message 是一个字符串，作为输出的前缀。
- first_n 表示打印输出结果的步骤数；如果该值为负，则只要执行路径，就会一直输出该值。
- summarize 表示从张量中输出的元素数量；默认情况下，仅输出三个元素。

可以在 Jupyter 笔记本 ch-18_TensorFlow_Debugging 中获得相关的代码。

下面来修改之前创建的 MNIST MLP 模型，并添加 print 语句：

```
model = tf.Print(input_=model,
                 data=[tf.argmax(model,1)],
                 message='y_hat=',
                 summarize=10,
                 first_n=5
                 )
```

当运行代码时，在 Jupyter 的控制台中获得以下内容：

```
I tensorflow/core/kernels/logging_ops.cc:79] y_hat=[0 0 0 7 0 0 0 0 0 0...]
I tensorflow/core/kernels/logging_ops.cc:79] y_hat=[0 7 7 1 8 7 2 7 7 0...]
I tensorflow/core/kernels/logging_ops.cc:79] y_hat=[4 8 0 6 1 8 1 0 7 0...]
I tensorflow/core/kernels/logging_ops.cc:79] y_hat=[0 0 1 0 0 0 0 5 7 5...]
I tensorflow/core/kernels/logging_ops.cc:79] y_hat=[9 2 2 8 8 6 6 1 7 7...]
```

使用 tf.Print( ) 的唯一缺点是该函数只提供有限的格式化功能。

18.3 使用 tf.Assert( ) 断言条件

调试 TensorFlow 模型的另一种方法是插入条件断言。tf.Assert( ) 函数接受一个条件，如果条件为 false，则输出给定张量的列表并抛出 tf.errors.InvalidArgumentError。

1）tf.Assert( ) 函数的定义为：

```
tf.Assert(
    condition,
    data,
    summarize=None,
    name=None
)
```

2）断言操作不像 tf.Print( ) 函数那样落在计算图的路径中。为了确保执行 tf.Assert( ) 操作，需要将其添加到依赖项中。例如，定义一个断言来检查所有输入是否为正：

```
assert_op = tf.Assert(tf.reduce_all(tf.greater_equal(x,0)),[x])
```

3）在定义模型时，将 assert_op 添加到依赖项，如下所示：

```
with tf.control_dependencies([assert_op]):
    # x是输入层
    layer = x
    # 添加隐藏层
    for i in range(num_layers):
        layer = tf.nn.relu(tf.matmul(layer, w[i]) + b[i])
    # 添加输出层
    layer = tf.matmul(layer, w[num_layers]) + b[num_layers]
```

4）为了测试这段代码，在第 5 迭代周期之后引入了一个异常（impurity），如下所示：

```
if epoch > 5:
    X_batch = np.copy(X_batch)
    X_batch[0,0]=-2
```

5）代码在前 5 个迭代周期时运行良好，但接下来就会抛出错误：

```
epoch: 0000    loss = 6.975991
epoch: 0001    loss = 2.246228
epoch: 0002    loss = 1.924571
epoch: 0003    loss = 1.745509
epoch: 0004    loss = 1.616791
epoch: 0005    loss = 1.520804

-----------------------------------------------------------------
InvalidArgumentError              Traceback (most recent call last)
...
InvalidArgumentError: assertion failed: [[-2 0 0]...]
...
```

除了 tf.Assert( ) 函数（它可以采用任何有效条件表达式）之外，TensorFlow 还提供以下断言操作，这些操作能检查具体条件，并且有简单的语法：

- assert_equal
- assert_greater
- assert_greater_equal
- assert_integer
- assert_less
- assert_less_equal
- assert_negative
- assert_none_equal

- assert_non_negative
- assert_non_positive
- assert_positive
- assert_proper_iterable
- assert_rank
- assert_rank_at_least
- assert_rank_in
- assert_same_float_dtype
- assert_scalar
- assert_type
- assert_variables_initialized

前面的断言示例也可以写成如下形式：

```
assert_op = tf.assert_greater_equal(x,0)
```

18.4 使用 TensorFlow 调试器（tfdbg）进行调试

TensorFlow 调试器（tfdbg）与其他流行的调试器（如 pdb 和 gdb）在工作方式上相同。要使用调试器，通常需要进行如下操作：

1）在代码中要中断并查看变量的位置设置断点；
2）在调试模式下运行代码；
3）当代码运行到断点处时，查看变量，然后继续下一步。

一些调试器还允许您在代码执行时以交互方式观察变量，而不仅仅是在断点处：

1）为了使用 tfdbg，首先导入所需的模块并将会话绑定在调试器包装器中：

```
from tensorflow.python import debug as tfd

with tfd.LocalCLIDebugWrapperSession(tf.Session()) as tfs:
```

2）接下来，将过滤器附加到会话对象上。附加过滤器与在其他调试器中设置断点的方式相同。例如，以下代码附加了一个 tfdbg.has_inf_or_nan 过滤器，如果有任何中间张量的值为 nan 或 inf，就会中断：

```
tfs.add_tensor_filter('has_inf_or_nan_filter', tfd.has_inf_or_nan)
```

3）当代码执行 tfs.run() 时，调试器将在控制台中启动调试器接口，可以在其中运行各种调试器命令监视张量值。

4）文件 ch-18_mnist_tfdbg.py 提供了使用 tfdbg 的代码。用 python3 执行这些代码会看到 tfdbg 控制台输出的结果（见图 18-1）。执行的具体代码如下：

```
python3 ch-18_mnist_tfdbg.py
```

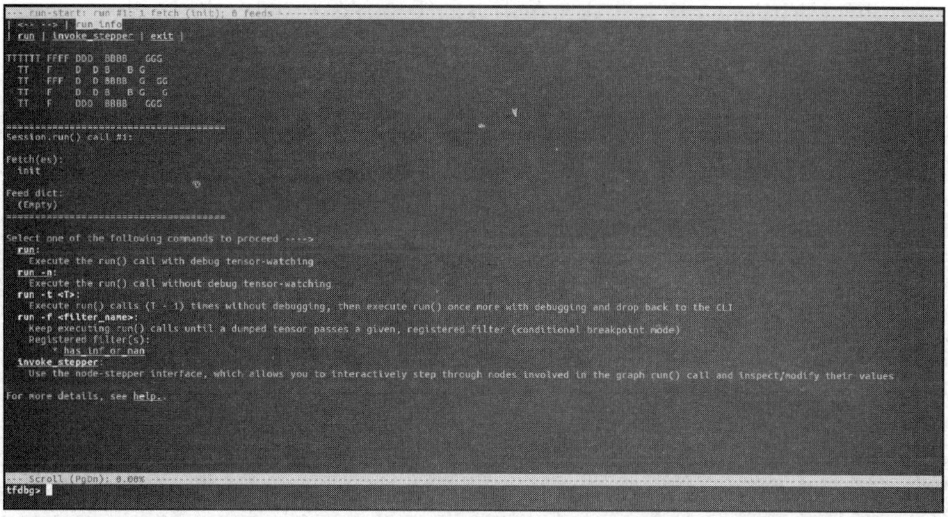

图 18-1

5）在 tfdbg> 提示符处输入命令 run -f has_inf_or_nan。由于使用 np.inf 值填充数据，代码在执行完第一个周期后会中断，如图 18-2 所示。

图 18-2

6）现在可以使用 tfdbg 控制台或可单击界面来检查各种张量的值。例如，查看其中一个梯度的值，如图 18-3 所示。

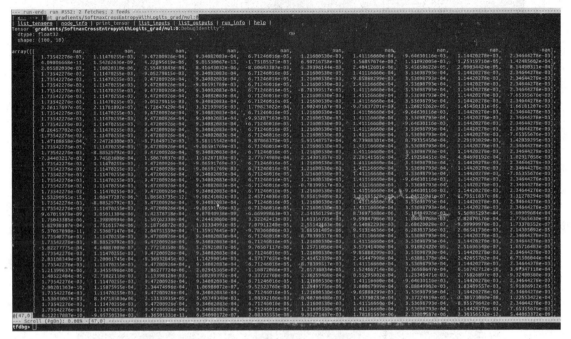

图 18-3

可以在以下链接中找到如何使用 tfdbg 控制台和如何检查变量的更多信息：https://www.tensorflow.org/programmers_guide/debugger。

18.5 总结

本章介绍了如何在 TensorFlow 中调试用于构建和训练模型的代码。通过本章了解到可以使用 tf.Session.run( ) 将张量作为 NumPy 数组来获取，在计算图中还可以通过添加 tf.Print( ) 操作来输出张量值。本章还介绍了在执行含有 tf.Assert( ) 操作和其他 tf.assert_* 操作的代码时，若某个条件不成立时如何抛出错误。本章最后介绍了 TensorFlow 调试器（tfdbg），能用 tfdbg 设置断点并观察张量值，就像在 Python 调试器（pdb）或 GNU 调试器（gdb）中调试代码一样。

本章的内容是一个新的里程碑，我们不希望在这里结束，相信这才刚刚开始，读者可进一步扩展和应用从本书中获得的知识与技能。

非常期待您分享经验，以及反馈建议。

附录
张量处理单元

张量处理单元（TPU）是**专用集成电路（ASIC）**，它是针对深度神经网络的计算要求而优化的硬件电路。TPU 基于**复杂指令集计算机**（Complex Instruction Set Computer，CISC）指令集，该指令集实现了用于训练深度神经网络的高级指令。TPU 架构的核心在于优化矩阵运算的脉动（systolic）阵列。TPU 架构如图 A-1 所示。

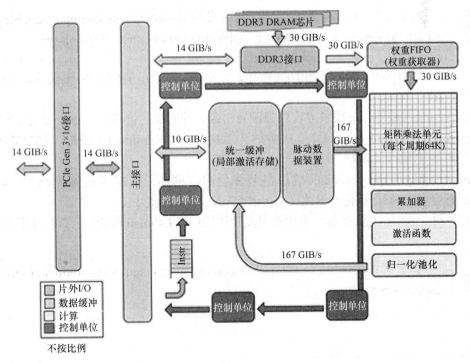

图 A-1

图片来自：https://cloud.google.com/blog/big-data/2017/05/images/149454602921110/tpu-15.png。

TensorFlow 提供了一个编译器和软件堆栈，可将 API 调用从 TensorFlow 计算图转换为 TPU 指令。图 A-2 描述了在 TPU 堆栈顶部运行的 TensorFlow 模型的体系结构。

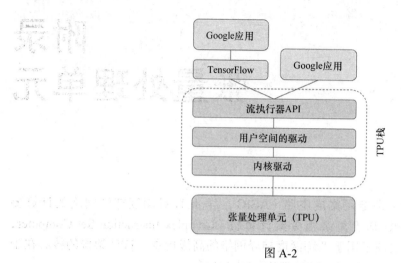

图 A-2

图片来自：https://cloud.google.com/blog/big-data/2017/05/images/149454602921110/tpu-2.png。

有关 TPU 架构的更多信息，请参考博客：https://cloud.google.com/blog/big-data/2017/05/an-in-depth-look-at-googles-first-tensor-processing-unit-tpu。

TPU 的 TensorFlow API 位于 tf.contrib.tpu 模块中，为了在 TPU 上构建模型，使用以下 3 个 TPU 特定的 TensorFlow 模块：

- tpu_config：tpu_config 模块允许创建配置对象，该对象包含即将运行模型的主机信息。
- tpu_estimator：tpu_estimator 模块将估计器封装在 TPUEstimator 类中。创建此类对象是为了在 TPU 上运行估算器。
- tpu_optimizer：tpu_optimizer 模块包装优化器。例如，在以下示例代码中，将 tpu_optimizer 模块中的 SGD 优化器封装在 CrossShardOptimizer 类中。

例如，对于 MNIST 数据集，下面的代码可使用 TF Estimator API 在 TPU 上构建 CNN 模型：

以下代码改编自：https://github.com/tensorflow/tpu-demos/blob/master/cloud_tpu/models/mnist/mnist.py。

```
import tensorflow as tf

from tensorflow.contrib.tpu.python.tpu import tpu_config
from tensorflow.contrib.tpu.python.tpu import tpu_estimator
from tensorflow.contrib.tpu.python.tpu import tpu_optimizer

learning_rate = 0.01
batch_size = 128
```

```python
    def metric_fn(labels, logits):
        predictions = tf.argmax(logits, 1)
        return {
            "accuracy": tf.metrics.precision(
                labels=labels, predictions=predictions),
        }

    def model_fn(features, labels, mode):
        if mode == tf.estimator.ModeKeys.PREDICT:
            raise RuntimeError("mode {} is not supported yet".format(mode))

        input_layer = tf.reshape(features, [-1, 28, 28, 1])
        conv1 = tf.layers.conv2d(
            inputs=input_layer,
            filters=32,
            kernel_size=[5, 5],
            padding="same",
            activation=tf.nn.relu)
        pool1 = tf.layers.max_pooling2d(inputs=conv1, pool_size=[2, 2],
                                        strides=2)
        conv2 = tf.layers.conv2d(
            inputs=pool1,
            filters=64,
            kernel_size=[5, 5],
            padding="same",
            activation=tf.nn.relu)
        pool2 = tf.layers.max_pooling2d(inputs=conv2, pool_size=[2, 2],
                                        strides=2)
        pool2_flat = tf.reshape(pool2, [-1, 7 * 7 * 64])
        dense = tf.layers.dense(inputs=pool2_flat, units=128,
                                activation=tf.nn.relu)
        dropout = tf.layers.dropout(
            inputs=dense, rate=0.4,
            training=mode == tf.estimator.ModeKeys.TRAIN)
        logits = tf.layers.dense(inputs=dropout, units=10)
        onehot_labels = tf.one_hot(indices=tf.cast(labels, tf.int32), depth=10)

        loss = tf.losses.softmax_cross_entropy(
            onehot_labels=onehot_labels, logits=logits)

        if mode == tf.estimator.ModeKeys.EVAL:
            return tpu_estimator.TPUEstimatorSpec(
                mode=mode,
                loss=loss,
                eval_metrics=(metric_fn, [labels, logits]))

        # 训练
        decaying_learning_rate = tf.train.exponential_decay(learning_rate,
                                                tf.train.get_global_step(),
                                                100000, 0.96)

        optimizer = tpu_optimizer.CrossShardOptimizer(
                tf.train.GradientDescentOptimizer(
                    learning_rate=decaying_learning_rate))
```

```python
        train_op = optimizer.minimize(loss,
            global_step=tf.train.get_global_step())
        return tpu_estimator.TPUEstimatorSpec(mode=mode,
            loss=loss, train_op=train_op)

def get_input_fn(filename):
    def input_fn(params):
        batch_size = params["batch_size"]

        def parser(serialized_example):
            features = tf.parse_single_example(
                serialized_example,
                features={
                    "image_raw": tf.FixedLenFeature([], tf.string),
                    "label": tf.FixedLenFeature([], tf.int64),
                })
            image = tf.decode_raw(features["image_raw"], tf.uint8)
            image.set_shape([28 * 28])
            image = tf.cast(image, tf.float32) * (1. / 255) - 0.5
            label = tf.cast(features["label"], tf.int32)
            return image, label

        dataset = tf.data.TFRecordDataset(
            filename, buffer_size=FLAGS.dataset_reader_buffer_size)
        dataset = dataset.map(parser).cache().repeat()
        dataset = dataset.apply(
            tf.contrib.data.batch_and_drop_remainder(batch_size))
        images, labels = dataset.make_one_shot_iterator().get_next()
        return images, labels
    return input_fn

# TPU 配置

master = 'local'  # TPU实例的 URL
model_dir = '/home/armando/models/mnist'
n_iterations = 50    # 每个TPU训练循环的迭代次数
n_shards = 8      # TPU芯片数量

run_config = tpu_config.RunConfig(
    master=master,
    evaluation_master=master,
    model_dir=model_dir,
    session_config=tf.ConfigProto(
        allow_soft_placement=True,
        log_device_placement=True
    ),
    tpu_config=tpu_config.TPUConfig(n_iterations,
                                     n_shards
    )
)

estimator = tpu_estimator.TPUEstimator(
```

```
            model_fn=model_fn,
            use_tpu=True,
            train_batch_size=batch_size,
            eval_batch_size=batch_size,
            config=run_config)

train_file = '/home/armando/datasets/mnist/train'  # 输入数据文件
train_steps = 1000  # 训练的步数

estimator.train(input_fn=get_input_fn(train_file),
                max_steps=train_steps
                )

eval_file = '/home/armando/datasets/mnist/test'  # 测试数据文件
eval_steps = 10

estimator.evaluate(input_fn=get_input_fn(eval_file),
                   steps=eval_steps
                   )
```

更多有关在 TPU 上构建模型的示例，请参见链接：https://github.com/tensorflow/tpu-demos。